AN INTRODUCTION TO
RADIO DXING

D1827458

ALSO BY THE SAME AUTHOR

AN INTRODUCTION TO
RADIO DXING

by
R.A. PENFOLD

BERNARD BABANI (publishing) LTD
THE GRAMPIANS
SHEPHERDS BUSH ROAD
LONDON W6 7NF
ENGLAND

Although every care has been taken with the preparation of this book, the author or publishers will not be held responsible in any way for any errors that might occur.

ISBN 0 85934 066 X

© 1981 BERNARD BABANI (publishing) LTD

First Published — September 1981

British Library Cataloguing in Publication Data
Penfold, R.A.
    An introduction to radio dxing.
    1. Radio — Receivers and reception —
    Amateurs' manuals
    I. Title
    621. 3841'1   TK 6563

Printed and bound in Great Britain by Cox & Wyman Ltd, Reading

PREFACE

Radio DXing (the reception of difficult or distant radio stations) has been a popular pastime for many years now, and is probably more popular now than ever before. While being a worthwhile and interesting hobby, the technicalities can be rather bewildering for the newcomer to the hobby, as can the choice of suitable equipment, and so actually getting started can prove difficult. You can simply buy a good quality communications receiver, connect a short length of wire to the aerial terminal, and start listening. However, you are far more likely to obtain good results if you know where on the dial to look for the various types of transmission, and when to look. Some knowledge of receiving equipment and techniques could also help to prevent the purchase of unsuitable or unnecessarily expensive equipment.

The purpose of this book then, is to introduce the various forms of radio DXing to beginners, with advice on suitable equipment and the techniques employed when using that equipment. For those interested in electronics construction a number of useful accessories will also be described. The book is divided into two main sections; one devoted to amateur band reception, and the other covering broadcast band reception. Most DXers eventually specialise in one or other of these, often ignoring the other completely. While this may seem a little strange, the two types of reception and the techniques involved are very different in many respects, so that one often appeals to the individual DXer far more than the other.

Initially it is difficult; probably impossible, to know which will be of most interest to you, without trying your hand at both. Thus it is advisable for the beginner to obtain a general coverage receiver that permits reception of all the short wave broadcast and amateur bands. In fact a general coverage receiver which has medium wave plus full short wave coverage, together with the usual facilities one would expect to find in a communications receiver, is suitable for virtually any type of radio DXing. Even VHF and UHF bands can be covered with the addition of a suitable converter.

Although we will be primarily concerned with short wave DXing, we will also consider DXing on the medium wave band and on VHF.

Readers Notes

(i) Please note that further to the World Administrative Radio Conference (WARC) of the International Telecommunications Union (ITU) held in Geneva at the end of 1979, certain changes were proposed to the frequency spans of some amateur and commercial broadcast bands, which should come into effect in January 1982. To ensure that this book is as up to date as possible, these new frequency limits have been used in the various tables included in the text.

(ii) The limits of the frequency spans of some amateur and commercial broadcast bands in countries other than the U.K. may vary slightly to those shown in this book.

CONTENTS

CHAPTER 1

Amateur Band DXing

Perhaps we should first make it quite clear what we mean by the amateur bands. These are parts of short wave spectrum which are set aside for private individuals who wish to operate their own transmitting and receiving stations. Some of these bands are exclusively for amateur use, while others are shared with other services. In order to obtain a licence to operate such a station it is necessary to show that you are competent to do so, and in the U.K. this involves passing an exam and a morse test. However, here we are only concerned with the reception of amateur transmissions, and no licence is necessary for this.

There are six short wave amateur bands in use, and these cover the frequency ranges quoted below:—

160 Metres (Topband)	1.800 MHz to 1.850 MHz
80 Metres	3.500 MHz to 3.800 MHz
40 Metres	7.000 MHz to 7.100 MHz
20 Metres	14.000 MHz to 14.350 MHz
15 Metres	21.000 MHz to 21.450 MHz
10 Metres	28.000 MHz to 29.700 MHz

In some countries (including the UK) the 160 metre band is extended to 2.0MHz, and in some countries (such as the USA) the 80 metre band extends to 4.0MHz and the 40 metre band is extended to 7.3 MHz. There are three new amateur bands that will be introduced in the future, but at the time of writing it is obviously not possible to say when these will be in full use. Anyway, the proposed bands are all very narrow, and have the following frequency ranges:—

29.5 (30) Metres	10.100 MHz to 10.150 MHz
16.5 (17) Metres	18.068 MHz to 18.168 MHz
12 Metres	24.890 MHz to 24.990 MHz

Requirements

We will now look at what is required of a receiver in order to give good results on the amateur bands. One obvious requirement is that it should cover as many of these bands as possible, and most general coverage sets include the entire short wave frequency spectrum and therefore cover all the bands. However, some of the older sets that are available on the second hand market do not cover the higher frequency bands, and a few are without coverage of 160 metres. This reduces their usefulness as far as the amateur band listener is concerned, but this is usually reflected in a low price tag which makes such a set attractive if only limited funds are available.

There are a few sets available which cover just the amateur bands, and these normally give coverage of all six current short wave amateur bands (plus the 2 metre VHF amateur band in some cases), but suprisingly a few of these sets omit 160 metres and (or) full coverage of the 10 metre band. The advantage of sets of this type is the excellent bandspread that they provide. In other words, the full length (or practically the full length) of each tuning scale is devoted to just one amateur band, and it will probably take many turns of the tuning dial to tune from one end of the band to the other. This usually gives such sets high calibration accuracy and makes precise tuning comparitively easy.

A similar degree of calibration accuracy and ease of tuning is achieved in some general coverage receivers by the use of a separate bandspread control and tuning dial. Here the main tuning control is adjusted to one end of the band to be covered, and then the bandspread dial is used to tune over the band and the bandspread dial indicates the reception frequency. This gives the advantage of general coverage plus good amateur band and (or) broadcast band bandspread, and is a system which I have found to work well in practice.

A number of modern sets use a system whereby general coverage is achieved by splitting up the frequency coverage into a number of fairly narrow tuning ranges (often 1MHz or 2MHz wide). This is another system which works well in practice giving both general coverage and good bandspread. Sets of this type often have digital readout of the reception frequency, giving extremely good calibration accuracy.

There are purely mechanical ways of obtaining good bandspread and precise calibration on a general coverage set, but these systems do not seem to be popular these days. In truth it is probably not too important what form of bandspread is used provided it is good and it is there! A set without any form of bandspread is likely to be very difficult to use on the amateur bands (fortunately virtually every SW set has some form of bandspread, even if it is just in the form of a slow motion drive fitted to the main tuning drive, but a method which gives really good bandspread is to be preferred due to the precision required when tuning an amateur transmission, as we shall see shortly).

It is necessary to have a tuning mechanism that is well constructed and free from backlash, otherwise pecise tuning can become very difficult, making the set difficult to use. Fortunately most sets seem to be quite good in this respect, and I have yet to come across a set where backlash is a real problem.

Selectivity

This is undoubtedly very important for an amateur bands receiver these days as the amateur bands are mostly not very wide, and the number of licensed amateur stations is into seven figures! It is therefore quite normal for the bands to be crammed with stations, making it difficult to pick out just a single transmission. Simply stated, the selectivity of a receiver is a measure of how well it does or does not pick up just one station from a crowded band. Selectivity is of such fundamental importance that we shall consider it in some detail before proceeding further.

The tuning dial of a receiver is usually calibrated in frequency, and we often talk of a set being tuned to a particular frequency. It is also quite common for a transmitter to be quoted as operating at a particular frequency. However, a receiver does not have infinite selectivity and does not pick-up just a single frequency and reject signals at any other frequency. It is sensitive over a small band of frequencies centred around the frequency indicated by the tuning dial. Similarly, with most types of transmission the output from the transmitter is not just a single frequency, but covers a small band of frequencies.

Ideally the transmitter should not produce frequencies over a very wide "bandwidth", as this would obviously limit quite severely the number of transmissions that could be accommodated on the amateur bands at any one time. One problem with an ordinary amplitude modulation (AM) signal (the type of modulation used by MW, LW and SW broadcast stations) is that it does tend to occupy quite a large part of the band. An ordinary AM signal consists of a carrier wave, plus two sidebands which are produced by the modulation of the carrier wave. The sideband that is higher in frequency than the carrier wave is called the "upper sideband", and the one which is below the carrier frequency is called the "lower sideband". This is shown in Figure 1.

The sidebands are placed symetrically around the carrier wave, and the spacing from the carrier to each of the sideband products is equal to the audio input frequency that generated that product. For example, audio input signals at frequencies of 0.5kHz, 1kHz and 2.5kHz would cause sideband signals spaced 0.5kHz, 1kHz and 2.5kHz respectively above and below the carrier frequency. The strength of each sideband product is proportional to the strength of the audio input frequency that produced it. In a practical signal, of course, the frequencies in the sidebands would normally be very numerous and constantly changing.

In order to give the full audio bandwidth of 20kHz an AM transmitter must obviously occupy some 40kHz of the band in which it is operating (20kHz either side of the carrier frequency). This is unacceptably large for an amateur band transmitter, since, as an example, only five such transmission could be accommodated in the 200kHz (0.2MHz) of the 160 metre band (1.8 to 2.0MHz).

Fortunately the full audio bandwidth is not required for effective voice communication, and there is little loss of intelligibility if frequencies above about 3kHz are attenuated. Thus a bandwidth of only about 6kHz can be achieved. It is quite common for the lower audio frequencies to also be rolled off to some extent as well, since these do not contribute significantly to intelligibility, and can actually hinder it. Rolling off these frequencies avoids wasting some of the power

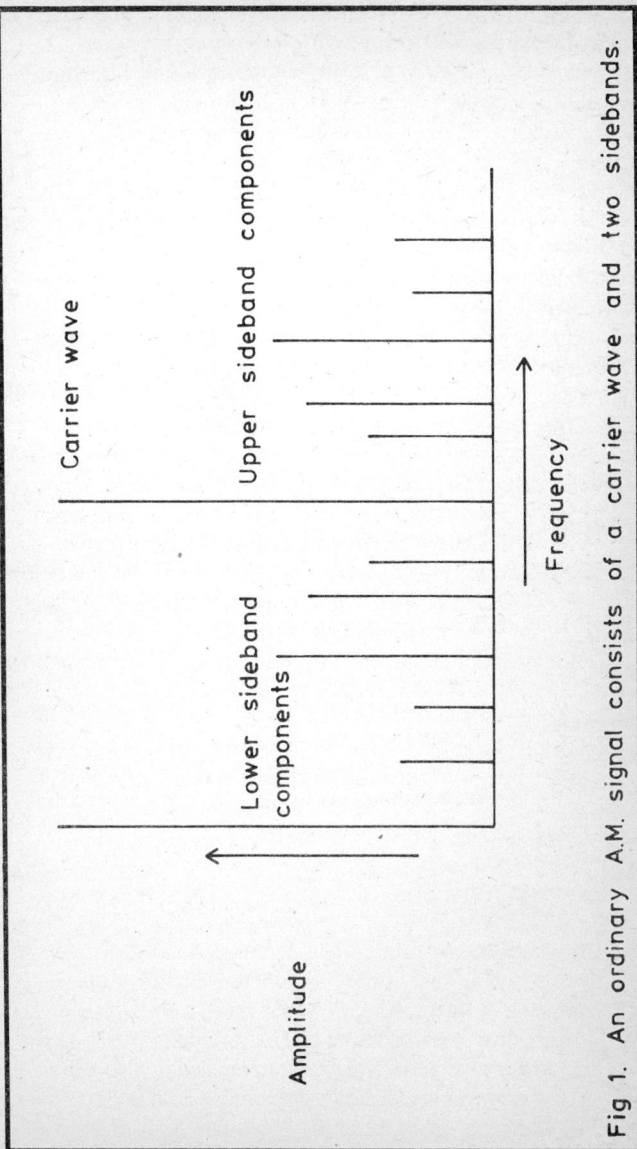

Fig 1. An ordinary A.M. signal consists of a carrier wave and two sidebands.

in the sidebands, and enables the sideband power to be concentrated where it will be most effective. This low frequency filtering does not give any reduction in the bandwidth of the signal.

Single Sideband

A further reduction in bandwidth can be achieved by eliminating one of the sidebands, since they both contain all the information needed to recover the original audio signal at the receiver. A small but worthwhile reduction in bandwidth (see Figure 2) can be achieved by also removing the carrier wave. Note that rolling off the lower audio frequencies does then have the effect of slightly reducing the bandwidth of the signal. In order to properly resolve the signal with the carrier wave suppressed it is necessary to reinsert the carrier at the receiver, and the set must have an oscillator which can be used to simulate the carrier and give a properly resolved signal. This oscillator is usually known as a BFO (beat frequency oscillator), but for the reception of this type of signal (usually known as single side band suppressed carrier, single sideband, or just SSB) it is more correctly called a carrier insertion oscillator or CIO for short.

There are two types of single sideband, upper sideband (USB) and lower sideband (LSB), and this merely refers to the sideband that is transmitted (the other one being suppressed of course). It is not merely of academic interest whether the signal being received is upper sideband or lower sideband, as we shall see shortly.

Receiver Bandwidth

So far we have only considered the bandwidth of the transmission; we will now consider bandwidth at the receiver. There is obviously no point in the receiver having a bandwidth which is larger than the bandwidth of the transmission, as this could easily lead to interference from transmissions close to but not overlapping the desired transmission (a form of interference known as adjacent channel interference). It would also give a noise level that was higher than necessary. Too little band-

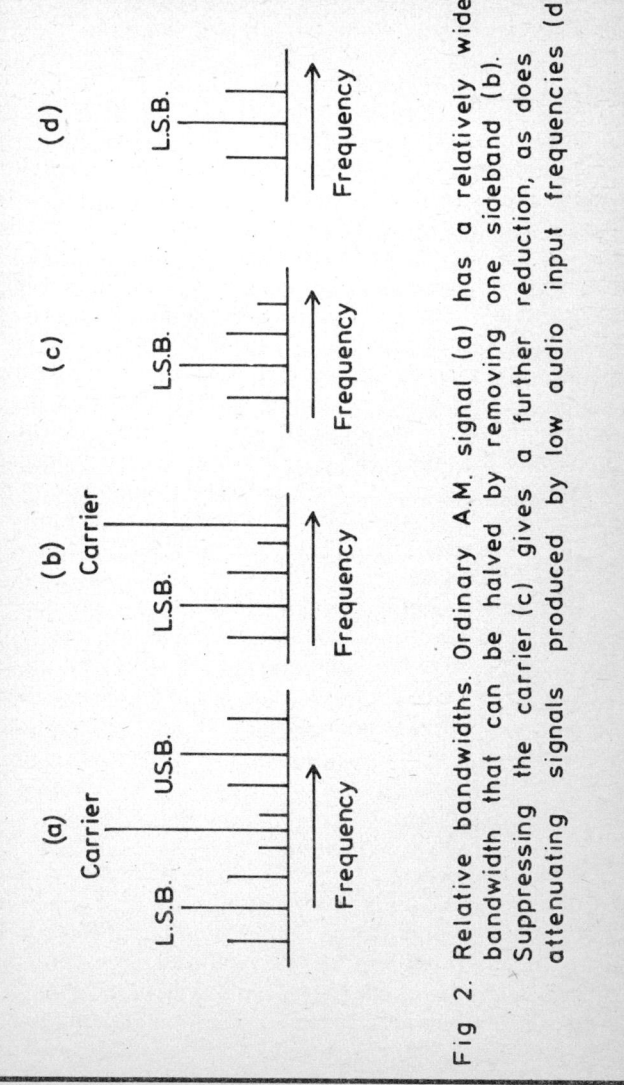

Fig 2. Relative bandwidths. Ordinary A.M. signal (a) has a relatively wide bandwidth that can be halved by removing one sideband (b). Suppressing the carrier (c) gives a further reduction, as does attenuating signals produced by low audio input frequencies (d).

width would have the effect of removing some of the frequency components of the transmission, in turn eliminating some of the audio output frequencies and imparing intelligibility. The bandwidth of the receiver therefore needs to be accurately matched to that of the transmission in order to give optimum results. This means that ideally the set should be capable of providing a different bandwidth for each transmission mode it is designed to resolve.

Ideally the response of the receiver should be as shown in Figure 3, with there being no variation in gain over the passband of the set, and infinite attenuation outside the passband. This would give the set a perfectly flat frequency response (and hence good audio output quality in this respect) plus total rejection of signals on adjacent channels. In practice such a response cannot be achieved, and most practical filters available today give a level of performance that is far removed from the ideal. However, there are some that do achieve a commendable level of performance, but as one would expect, these are generally the more expensive types.

Given that a receiver is not going to achieve the ideal response, we obviously want one that will come as close as possible. The normal way of expressing the efficiency of the filter used in a receiver is to give its "slope factor". This is the bandwidth at its −60dB point divided by the bandwidth at the −6dB points (occasionally the −40dB and −6dB bandwidths are used). The −6dB points are where tha gain of the set falls to half its peak value, and the −60dB points are where the gain has fallen by a factor of 1,000, incidentally. Thus a filter having the response shown in Figure 4 would have a slope factor of 2 or 2:1. Of course, the ideal slope factor is 1, and the closer to this figure the better the performance of the receiver and filter. In practice a slope factor of around 1.5 to 2 is considered very good.

Tuning an SSB Signal

Tuning in an SSB signal is not like tuning to an AM or FM broadcast station; the tuning must be pretty well spot on or the audio output will be completely unintelligible. As explained earlier, an oscillator in the receiver is used to

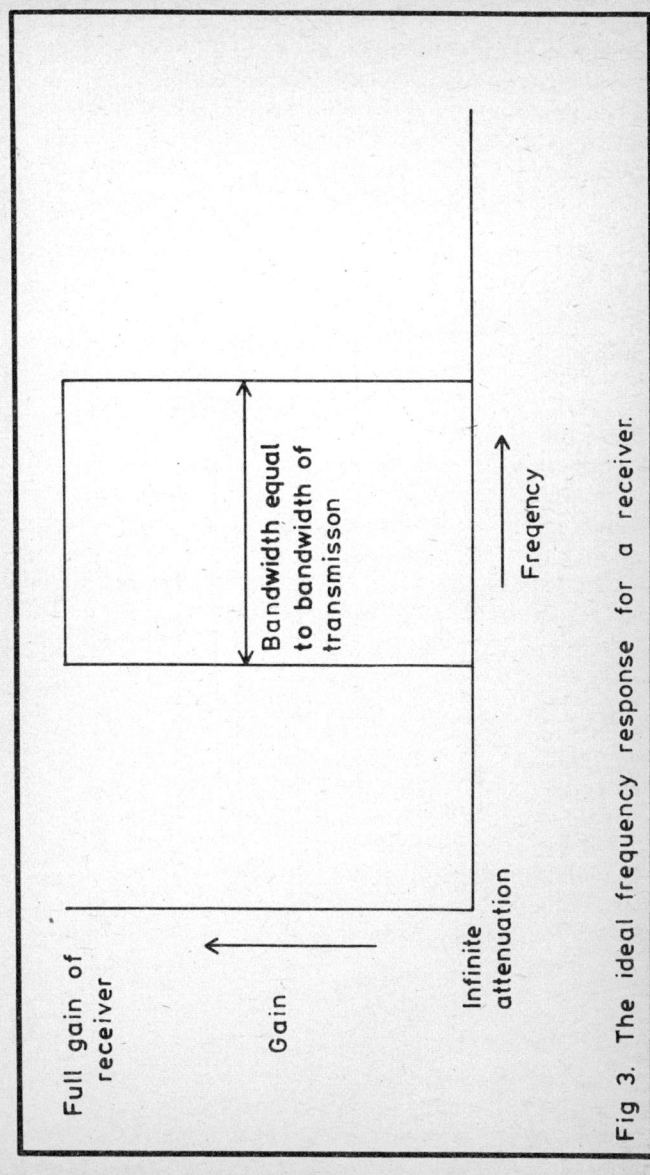

Fig 3. The ideal frequency response for a receiver.

9

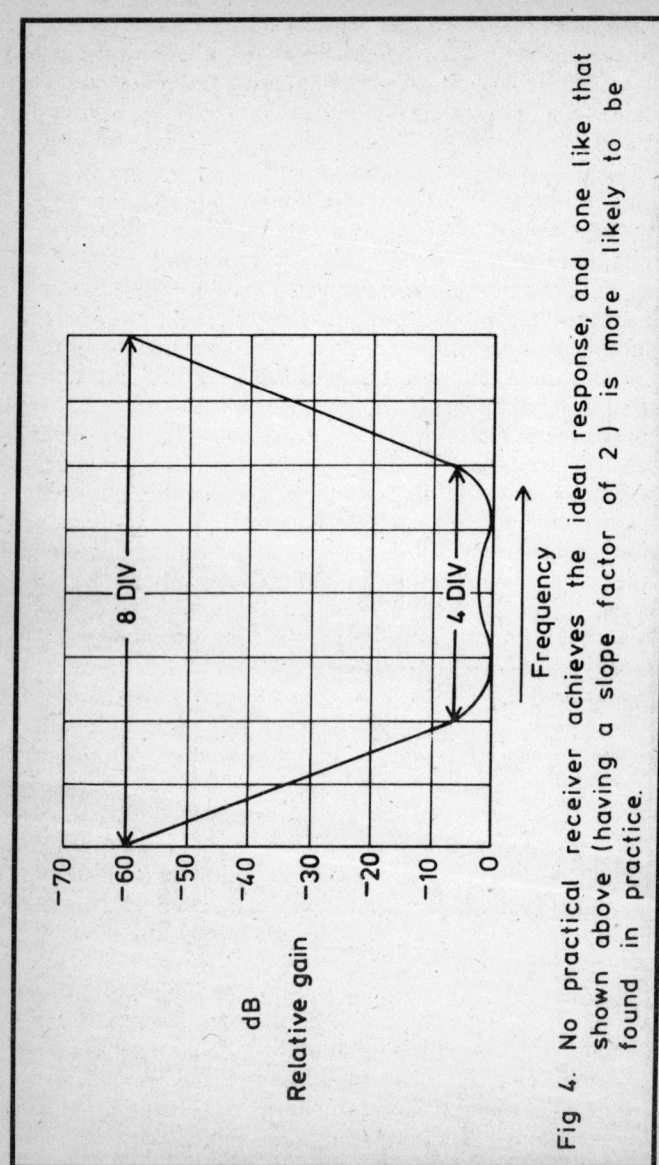

Fig 4. No practical receiver achieves the ideal response, and one like that shown above (having a slope factor of 2) is more likely to be found in practice.

simulate the suppressed carrier wave of the received transmission. Some sets have an oscillator that can be varied in frequency slightly, usually by up to just a few kHz either side of the central reception frequency. Other sets have two switched oscillator frequencies, one offset a few kHz above the central reception frequency, and the other offset by a similar amount in the other direction. The oscillator is offset in one direction to receive upper side band, and offset the other way for the reception of lower side band.

The reason for this is explained in Figure 5. Here we have a lower sideband signal correctly tuned so that it fits nicely into the passband of the receiver. If it was not removed, the carrier wave would be just off the high frequency end of the LSB signal components. In order to correctly resolve the signal the oscillator must obviously be tuned to the high frequency end of the receiver's passband, so that it effectively replaces the carrier wave. In the case of an upper sideband signal the carrier wave, if it had not been suppressed, would appear at the low frequency end of the receiver's passband, and this is where the oscillator must be tuned in order to properly resolve the signal.

If the oscillator is not quite in the right place it may still be possible to resolve the signal, but some of the sideband components may be removed by the receiver since they will be outside its passband (as shown in Figure 6), and this will result in a loss of high or low audio frequencies from the output.

If the oscillator is offset in the wrong direction the signal will seem to become very weak as it is tuned in, and may even seem to dissapear altogether! This is simply because the signal will be outside the receiver's passband if the set is tuned to bring the oscillator to the position where the suppressed carrier would be (see Figure 7). It would of course be possible to tune the station in accurately even with the oscillator at the wrong side of the passband, but this would simply give a totally "scrambled" and unintelligible output. The audio output frequencies obtained are equal to the difference between the oscillator frequency and each of the sideband component frequencies. With the oscillator at right position this gives an audio output which is the same as the audio input to the transmitter (ignoring any frequencies that are lost due to

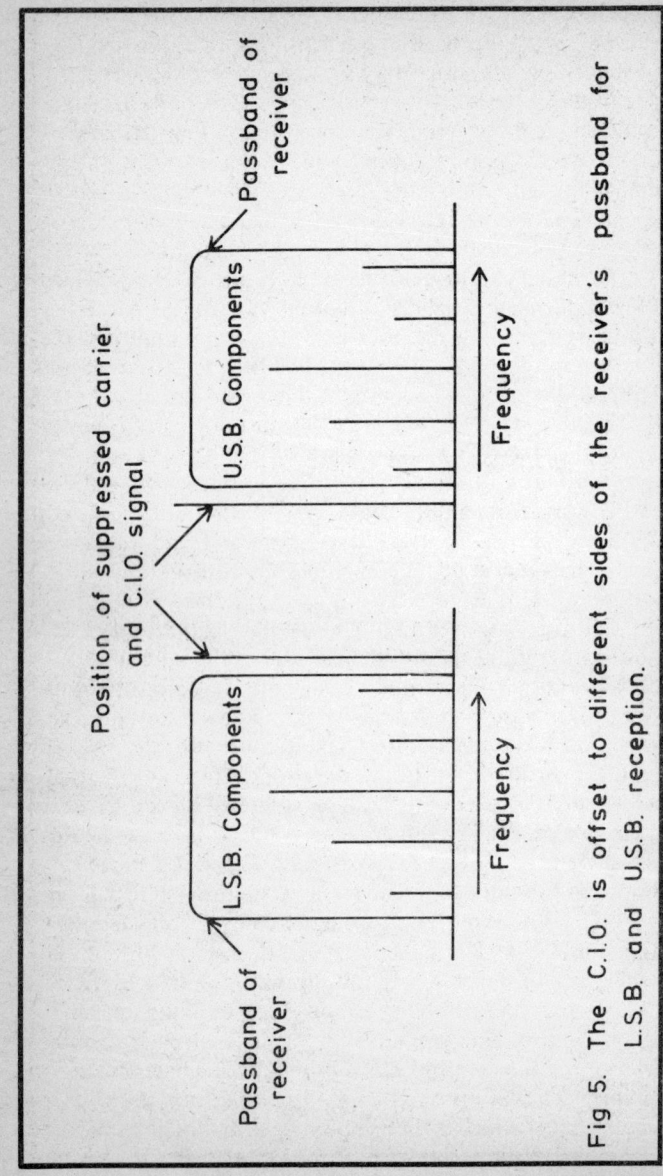

Fig 5. The C.I.O. is offset to different sides of the receiver's passband for L.S.B. and U.S.B. reception.

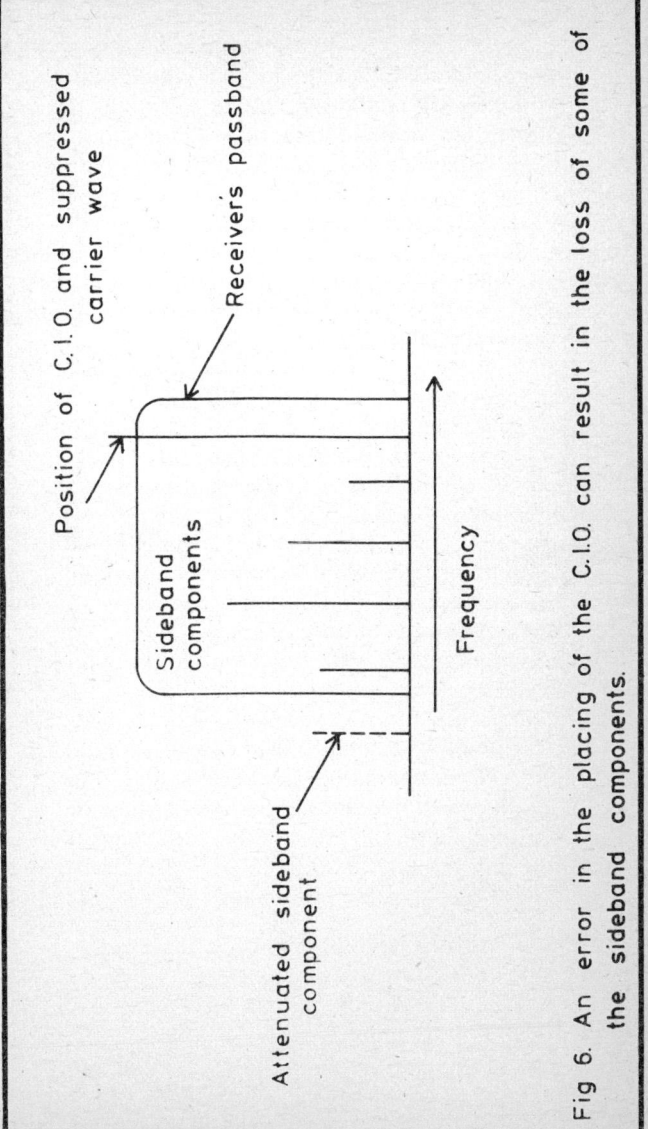

Fig 6. An error in the placing of the C.I.O. can result in the loss of some of the sideband components.

filtering in the system). With the oscillator on the wrong side of the signal the audio frequencies at the output of the receiver become changed completely, and undergo a form of inversion. The sideband components which should be closest to the oscillator so that they produce the low audio output frequencies become the furthest removed, and produce high frequencies! Similarly, the frequencies that are closest to the oscillator frequency and produce the lower audio output frequencies should in fact be the signals that are furthest away, and producing the high audio output frequencies. Not suprisingly, this gives an audio output which is not understandable, and does not even sound very much like a human voice. This effect is demonstrated in Figure 7.

As many readers will have gathered by now, tuning in an SSB signal is not just a matter of roughly tuning the set for maximum signal strength, with the exact setting of the tuning control not being especially critical. The tuning must be spot on or the audio output signal will be drastically altered, making it difficult or even impossible to comprehend. A slight tuning error with sideband signal set too far away from the oscillator results in all the audio frequencies being raised slightly (sometimes called the Donald Duck effect as it gives a sound rather reminiscent of the well known cartoon character). A slight tuning error in the other direction has the opposite effect with all the audio output frequencies being reduced somewhat in pitch.

Therefore, the set is carefully tuned until an audio output of the correct pitch is produced. The receiver must be set correctly to the upper sideband or lower sideband mode, as appropriate, or the signal will drop greatly in strength as the set is adjusted to give the correct audio output. In general lower sideband is used on the low frequency (LF) amateur bands (160, 80 and 40 metres), while upper sideband is used on the high frequency (HF) amateur bands (20, 15 and 10 metres). Not all stations keep to this convention, but exceptions are few and far between.

If a set having a variable BFO or CIO frequency control is being used, the optimum setting for this can be found by first tuning the set for maximum signal strength (it may be easier to do this with the set adjusted to the ordinary AM mode), and

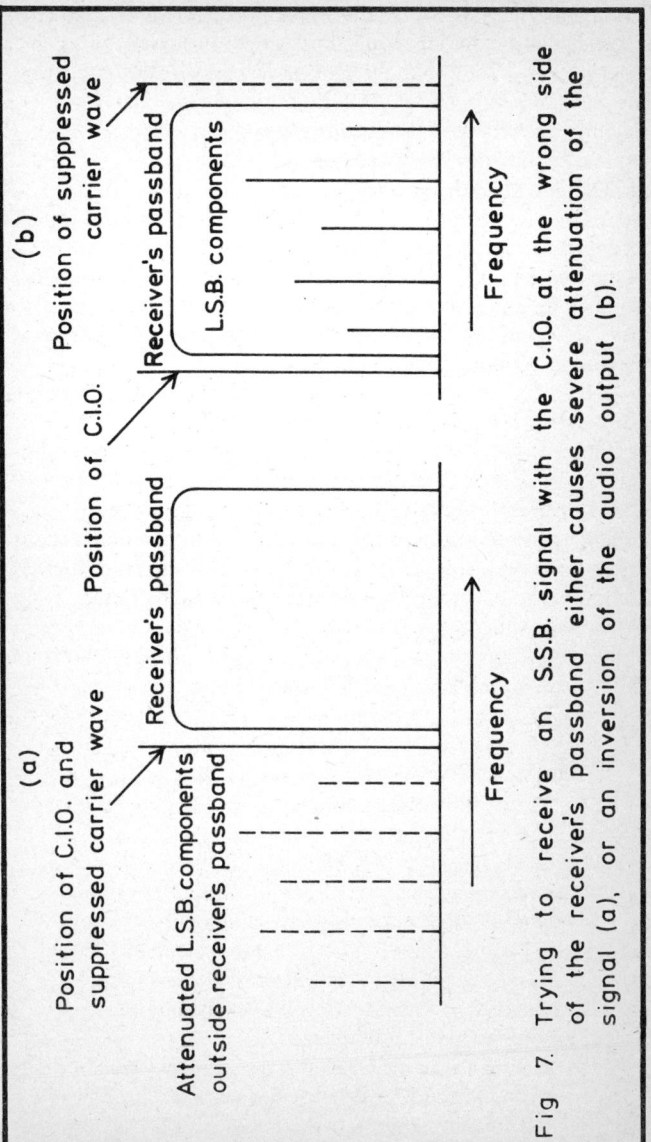

Fig 7. Trying to receive an S.S.B. signal with the C.I.O. at the wrong side of the receiver's passband either causes severe attenuation of the signal (a), or an inversion of the audio output (b).

15

then adjusting the BFO or CIO frequency control for the correct audio pitch. This adjustment must, of course, be carried out on both upper sideband and lower sideband signals in order to find the best settings for each mode.

Ordinary AM is still used to some extent on the short wave amateur bands, but the advantages of SSB have led to a decline in AM to the point where it is something of a rarity. Apart from the narrower bandwidth of SSB, it also has the advantage of not wasting transmission power on the carrier wave, which is a non-essential part of the transmission. Thus for a given input power an SSB transmitter will give a much more effective output. There are other advantages, and an SSB signal is, for example, less affected by certain atmospheric effects which cause fading.

CW Signals

Apart from AM and SSB, the only other mode of transmission which is used on the short wave amateur bands to any large extent is CW (continuous wave) which is used to send Morse code signals. A CW signal consists of just a single frequency which is switched on and off using a morse key. An ordinary AM receiver will not normally give an audio output from a signal of this type, but an SSB receiver will do so.

If we consider the situation depicted in Figure 8, the oscillator and the CW signal will react together to produce an audio output signal having a frequency equal to the difference in the frequency of the two signals. It is just the same as if the CW signal was a sideband component. It does not really matter where within the receiver's passband the CW signal is placed, this is really just a matter of adjusting the tuning control for the audio output tone that the operator finds most acceptable. It does not really matter whether the set is set to the upper sideband or the lower sideband mode, a suitable audio output will obviously be produced either way. It is important that the CW signal is tuned into the passband of the receiver, and is not tuned to the wrong side of the oscillator. This would give an audio output, but the set would have an apparent lack of sensitivity.

Although I stated above that it does not really matter where

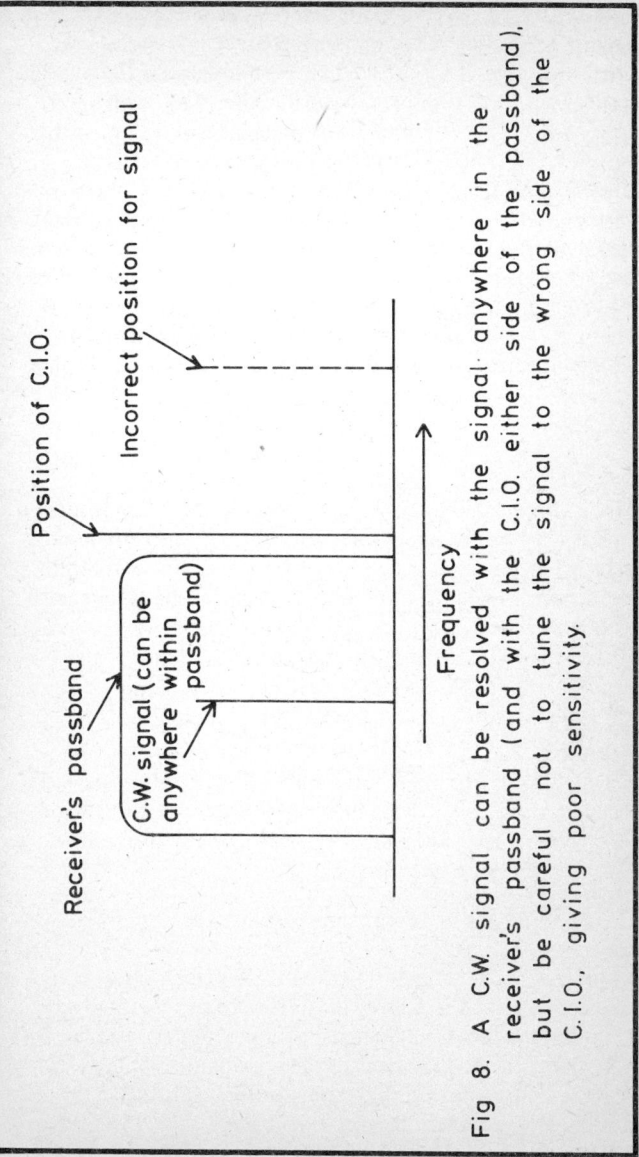

Fig 8. A C.W. signal can be resolved with the signal anywhere in the receiver's passband (and with the C.I.O. either side of the passband), but be careful not to tune the signal to the wrong side of the C.I.O., giving poor sensitivity.

Position of C.I.O.

Incorrect position for signal

Receiver's passband

C.W. signal (can be anywhere within passband)

Frequency

within the receiver's passband the CW signal is placed, it can sometimes be beneficial to adjust the tuning to minimise interference from a nearby signal, rather than tuning the receiver for the desired audio output frequency. Figure 9 shows how this reduction in interference can sometimes be achieved. This is where the tunable type of BFO or CIO is helpful, as with this type of control it is possible to first tune the set to minimise adjacent channel interference, and then adjust the BFO or CIO control to give an audio output of the desired pitch.

As will probably be apparent, a receiver having a bandwidth intended for AM and CW reception is not ideal for CW reception where a much narrower bandwidth is adequate, and reduces the likelyhood of adjacent channel interference. In theory an infinitely narrow bandwidth is acceptable for CW reception since just a single frequency is involved. Such a bandwidth is also ideal in that it would give an infinite number of CW channels and zero adjacent channel interference.

In practice though, an infinitely narrow bandwidth could not be achieved, and even if it could it would be impossible to locate a signal and keep the receiver accurately tuned to it. Receivers and transmitters both tend to drift slightly in frequency, and an inadequate bandwidth would therefore probably result in the signal constantly drifting out of the receiver's passband, making it necessary to try to relocate it each time.

A bandwidth of a few hundred Hz is probably about the minimum that can be used under practical operating conditions. A narrower bandwidth would probably be too difficult to use, and most of the time the inconvenience of such a bandwidth would not actually give any reduction in adjacent channel interference, and taken overall would not be advantageous.

Filters

If we now return to our original theme of the list of requirements for an amateur band short wave receiver, the ideal set would obviously have some form of variable selectivity, such as a set of filters to give optimum selectivity for each of the three main transmission modes (AM, SSB and CW). Also,

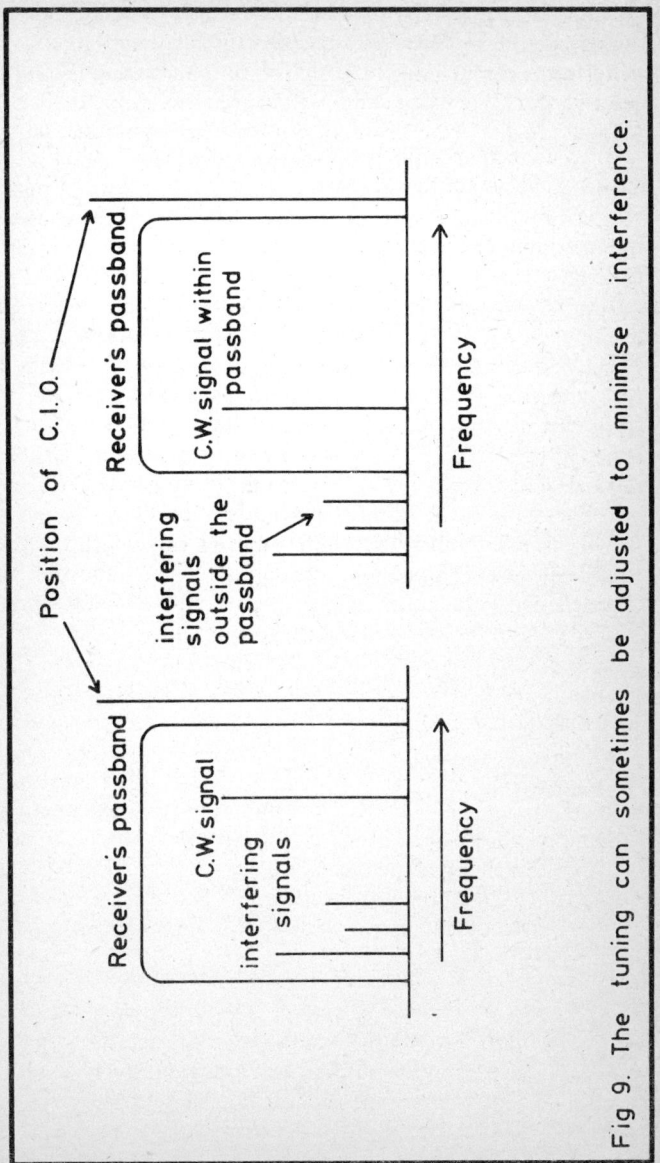

Fig 9. The tuning can sometimes be adjusted to minimise interference.

the set should have filters that have a good slope factor. Unfortunately, this is an expensive feature to incorporate in a receiver, and is something that is only likely to be found in sets selling at quite high prices.

It must be emphasised though, that although comprehensive and high performance filtering is a highly desirable asset in an amateur band short wave set, it is not an essential feature, and good results can be obtained using a set having a fixed and relatively wide bandwidth. Results on the amateur bands are determined largely by the skill of the operator; good results being possible with relatively simple equipment, and the best of equipment by no means guaranteeing excellent results. However, as one would expect, things are that much easier using high quality equipment, although you may prefer the challenge of using relatively simple equipment. This is something that one must decide for ones self.

A useful add-on circuit for CW reception with a set having a relatively wide bandwidth is an audio filter having a narrow passband. Units of this type simply connect between the headphones and the headphone output of the receiver, and provide a high level of rejection except over a small range of frequencies. In use the set is adjusted so that the CW note lies within this band of frequencies, hopefully leaving other signals outside the passband of the filter so that they are severely attenuated.

Audio Filter

The circuit diagram of an audio filter of this type is shown in Figure 10. L1 and C3 form a parallel tuned circuit having a resonant frequency of about 1.8kHz, although this can be changed slightly, if desired, by altering the value of tuning capacitor C3.

A parallel tuned circuit has a high impedance at or close to resonance, but the impedance reduces rapidly either side of resonance. Thus there is little voltage drop across series input resistor R1 at and around 1.8kHz, but at frequencies significantly removed from this figure the losses through R1 become quite high. This gives the required narrow passband, and the filter has a bandwidth of about 200Hz at the −6dB points

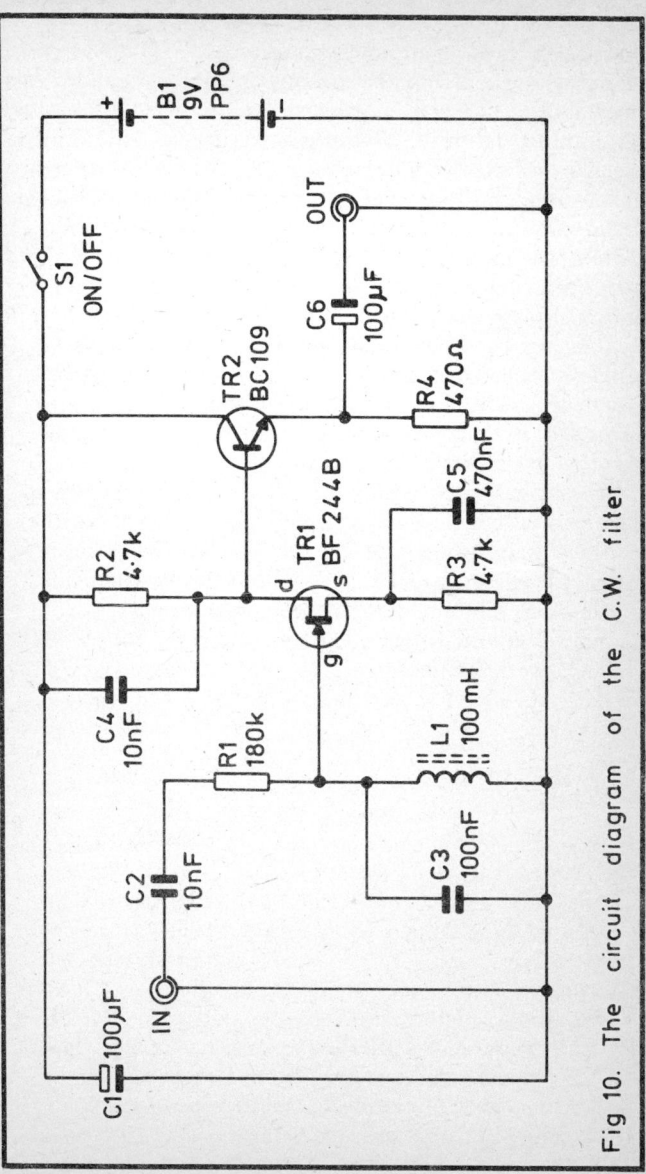

Fig 10. The circuit diagram of the C.W. filter

21

This is about the minimum that is reasonably easy to use in practice. C2 merely provides DC blocking at the input.

Tr1 is used as a common source amplifier, and this gives a small amount of voltage gain which compensates for the losses that occur through R1 at pass frequencies. L1 acts as the gate bias circuit for Tr1, and the very high input impedance of Tr1 ensures that there is no significant damping of the tuned circuit, and that a sharp response is obtained. The sharpness of the response is also helped by C5 which rolls off the low frequency response of the circuit, and C4 which gives increasing attenuation at the higher audio frequencies.

Tr2 is used as a simple common emitter output stage which is direct coupled to drain of Tr1. This gives the circuit a low output impedance so that it can drive high or medium impedance headphones without the output being excessively loaded. The unit also seems to work quite well with low impedance headphones, although these are considerably less than ideal in theory.

The only control is on/off switch S1. The voltage gain of the circuit is approximately unity, and so there should be no problems with excessive or inadequate gain when it is added into the system. The current consumption of the circuit is around 10mA.

Components: CW Filter, Figure 10

Resistors, all ¼ watt 5%

R1	180k	R2	4.7k
R3	4.7k	R4	470 ohms

Capacitors

C1	100 μF 10v	C2	10nF plastic foil
C3	100nF plastic foil	C4	10nF plastic foil
C5	470nF plastic foil	C6	100 μF 10v

Semiconductors

Tr1	BF244B	Tr2	BC109

Inductor

L1   100mH on ferrite pot core (MES)

Switch

S1   SPST subminature toggle type

Miscellaneous
PP6 size 9 volt battery, case, component panel, input and
output sockets, wire, solder, etc.

Facilities

We will now briefly consider some of the more popular
facilities found on short wave receivers, starting with AGC
(automatic gain control). This is a circuit which has no
significant effect when weak signal levels are being received,
but reduces the gain of the circuit when higher signal levels
are picked up. The higher the amplitude of the received
signals, the lower the gain of the set becomes.

This gives two beneficial effects: a wide range of input
levels gives a fairly constant audio output level, avoiding the
necessity to constantly adjust the gain of the set manually,
and stations that are prone to fading give a reasonably
consistent output level due to the stabilising effect of the
AGC. Of course, in cases where a station is quite weak and it
sometimes fades out to virtually nothing, the AGC circuit will
be of little or no help.

On some simple sets the AGC system does not operate in
the CW and SSB modes. The reason for this is simply that the
signal from the BFO or CIO would operate the AGC system,
reducing the sensitivity of the set. A simple way around this
problem is to disable the AGC circuitry. An alternative method
is to use a low level of BFO or CIO injection so that this
signal level is not large enough to produce a significant reduction
in sensitivity. The disadvantage of this system is that many
input signals will be large enough to swamp the BFO or CIO
signal, even though the AGC will to some extent counteract
this. On CW this does not really matter, and will simply result
in the audio output level being somewhat less than it might
otherwise be. On SSB there is both a loss of audio level and a
rather high distortion level, and in extreme cases the audio
output might be completely uncomprehendable. Of course,
even with the system where the AGC circuit is switched out
and a high level of BFO injection is used it is still possible for a
strong input signal to overload the CIO signal, and such a
signal could also overload the detector.

With sets of this type it is really necessary to have an RF gain control or an aerial attenuator control. The former permits the gain of the amplifier at the input of the set to be manually controlled, and the latter effectively enables the strength of the aerial signal supplied to the receiver to be controlled. The two controls are used in much the same way, and I have come across sets where an input attenuator control has been given the more familiar "RF Gain" legend. They can both be used to reduce signal strengths in the event of the receiver being overloaded.

The RF gain or aerial attenuator control can also be used to reduce the signal level in the early stages of the receiver if overloading of these stages causes severe cross modulation. The bandwidth of the early stages of a receiver is usually quite wide, with the narrow bandwidth required for good AM, CW and SSB reception being provided immediately after these input stages. Thus, although perhaps just one quite weak signal finds its way through the various stages of the set to produce a proper audio output, there may be numerous strong signals being processed by the input stages of the set. If these signals are strong enough they can cause serious distortion to be produced in these stages, producing new signals at many frequencies, including some that will be passed through the various stages of the set and will appear at the output. On the broadcast bands where there are a multitude of strong AM signals this can result in a strange effect where it sounds as if the set is picking up two stations at once. In fact what is happening is that the set is tuned to one station, and overloading at the front end of the receiver is producing new frequencies that are heard as the second station (which will be a powerful signal on a nearby channel). It is from this effect that the term "cross modulation" is derived. The unwanted modulation can usually be eliminated simply by backing off the RF gain or aerial attenuator control slightly so that the overloading ceases.

On the amateur bands there is much less use of ordinary AM as the mode of transmission, and cross modulation is less likely to manifest itself in this form. It is more likely to be heard as a higher than normal level of general noise, and if the RF gain or aerial attenuator control is backed off there should be a quite abrupt fall in the noise level at some point. It will probably be found that a number of weak signals can then be copied, where-as they were previously swamped in noise. Thus the strongest possible input signal does not always give the best results.

Ordinary transistors do not generally give very good cross modulation performance, although there are discrete and integrated circuit designs which do give commendable performance in this respect. Field effect transistors and valves usually give good cross modulation performance.

Product Detector

The more recent and more sophisticated communications receivers have a product detector for CW and SSB reception. Rather than simply injecting the BFO or CIO signal into the receiver at some point ahead of the AM detector, so that the two signals combine to produce an audio output from the detector, the amplified signal is taken to a separate detector (the product detector) where it is combined with the CIO or BFO signal to give an audio output. The amplified signal can still be applied to the AM detector (which normally also gives the signal which operates the AGC circuitry), or to the separate AGC signal detector if the circuit is of the type. There are other methods of obtaining the AGC control signal, such as deriving it from the audio output of the product detector. In either event, the obvious advantage of using a product detector is that it enables an effective AGC action to be provided during CW and SSB reception.

A good product detector only produces an output signal where an input signal has reacted with the CIO or BFO signal to produce an audio signal, and there is no significant audio output due to the various input components reacting with one another. This gives a "cleaner" and less distorted audio output signal.

Aerial Trimmer

Most receivers use a tuned circuit at the input, and this peaks sensitivity at the desired reception frequency while attenuating other frequencies. At least, it does if it is aligned properly with the other tuned circuits in the receiver. Unfortunately, the operating frequency of the input tuned circuit is affected to some extent by the type of aerial used with the set. Thus it is not possible to guarantee that this tuned circuit will be perfectly aligned under all operating conditions, even if the set has just been aligned and set up properly.

Receivers also tend to drop off in performance with the passing of time, mainly due to slight physical changes in the set causing a loss of alignment accuracy. This can be due to materials used in components altering slightly with the passing of time, giving a change in their electrical values, or it can simply be due to something like vibration causing the setting of a trimmer or core to change slightly.

An aerial trimmer enables the frequency of the aerial tuned circuit to be trimmed over a small range of frequencies, so that in the event of its alignment not being spot on it can be adjusted to resonate at the correct frequency and peak received signals.

"S" Meter

An "S" meter, or signals strength meter, is a useful accessory to have on a communications receiver. Meters of this type are usually calibrated in "S" units from 1 to 9, and above S9 in dB (usually up to a maximum of 40 or 60dB).

Really an "S" meter only provides a comparitive indication of signal strength, and this makes it useful when adjusting the set to peak received signals (the "S" meter then acting as a tuning meter or indicator). It can also show how changes in conditions cause variations in the strengths of received signals, and show which of two signals is the stronger (and give a rough idea of how great or small the difference in signal strengths happens to be).

It is important to realise though, that the actual "S" meter reading is virtually meaningless. Although it is often thought that the "S" meter reading shows the field strength of the received signal, this is not really the case. Two receivers set up at the same spot and tuned to the same station would probably produce two completely different "S" meter readings, even if both sets were of the same type! The reason for this is that the performance of receivers inevitably varies significantly from one set to another, even between two sets of the same type. Apart from the sensitivity of the set, the "S" meter reading also depends upon such things as the aerial used, the type of "S" meter circuit used in the set, and also how well or otherwise the set is adjusted. An "S" meter reading of say S7 therefore tells you very little, but does enable you to compare the strength of the signal against other signals, check to see if one aerial gives better results than another, and things of this nature.

Superhet System

Virtually all commercially produced communications receivers use some form of superheterodyne (superhet circuit). Figure 11 outlines the make up of a simple superhet receiver in block diagram form.

At the input there is usually an RF amplifier, although the aerial can be coupled direct to the frequency changer stage. The purpose of the frequency changer is to convert the incoming signal to the intermediate frequency (IF). Although this may seem to be a pointless exercise, it does in fact enable a very high level of performance to be obtained both in terms of sensitivity and selectivity.

The IF amplifier, detector, AGC, and audio circuitry form what is really a complete radio receiver, but one that is not tunable and has a fixed reception frequency (whatever IF is chosen). Since it operates on a fixed frequency it is easy to screen the tuned circuits from one another so as to avoid stray feedback which could cause instability; in fact it is easy to arrange all the circuitry so that good stability is obtained despite the high gain provided by the IF circuitry. Thus the set achieves good gain and sensitivity.

Perhaps of more importance, it is relatively easy to design a filter that gives a narrow bandwidth and good slope factor if it only needs to operate at a fixed frequency and does not need to be tunable. The reason for this is that there are crystal and ceramic resonators which act rather like ordinary L – C tuned circuits in the way they behave electrically, but they give a level of performance (Q) which is far higher than can be achieved with normal inductors and capacitors. Filters using

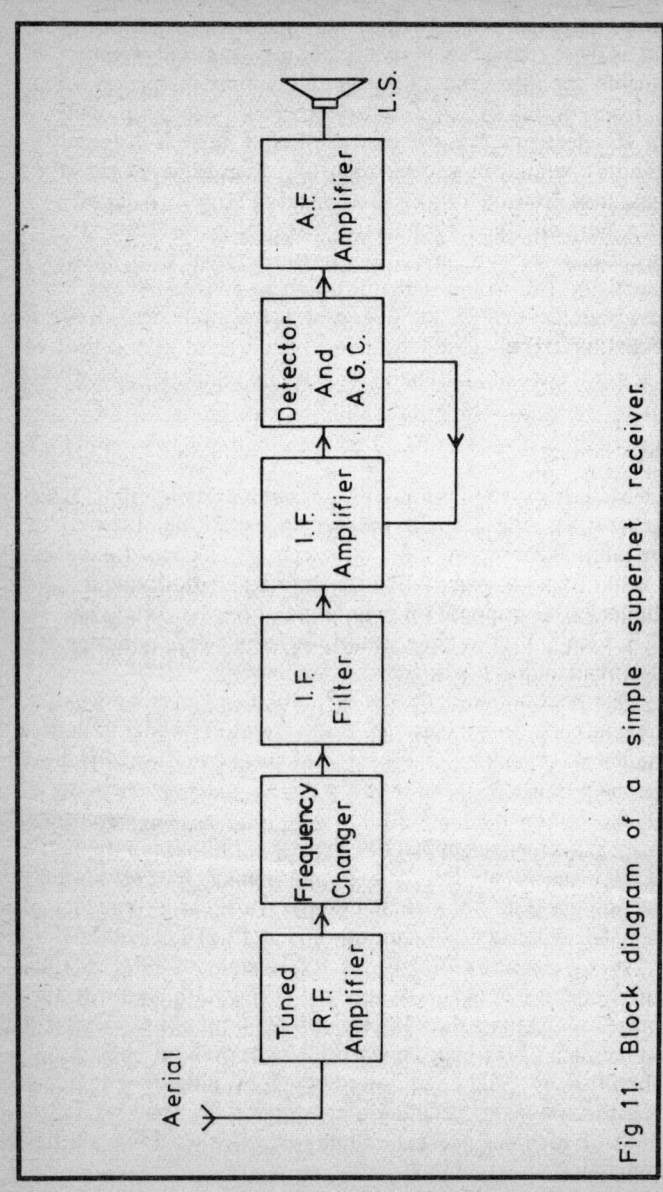

Fig 11. Block diagram of a simple superhet receiver.

28

these components can provide very high levels of performance, but as these resonators have a fixed frequency and are not tunable, the filters can only operate at a fixed frequency.

Even if ordinary L — C tuned circuits are used to provide the IF selectivity, it is still easier to have a number of tuned circuits operating at a fixed frequency, than to have each of these made tunable using a multiple gang tuning capacitor. Also, normally the intermediate frequency is lower than the reception frequency, and so the IF tuned circuits give better selectivity than would the same number of tuned circuits, having an identical Q, and operating at the input signal frequency. It is also relatively easy to obtain a high level of gain at the lower intermediate frequency than at the signal frequency.

Disadvantages

The superhet system is not without disadvantages, one of which is that it has at least two spurious responses. One of these is at the intermediate frequency, but as this is a constant frequency it is quite easy to place a filter in the signal path that will eliminate this response (in practice this filter, known as an "I.F. Trap", is often unnecessary, with the circuitry having a high inherent attenuation at this frequency).

The other response is more of a problem, and is known as the image response. This occurs at a frequency equal to double the IF away from the reception frequency, and is usually above the main response of the set. The tuned circuits at the input of the receiver are used to peak the main response and reject the image response, but in practice there is only limited attenuation of the image. If we consider a practical example, a set having the popular IF of 455kHz (0.455MHz) tuned to a frequency of 29.0MHz will have an image response at 29.910MHz (2 x 0.455 = 0.910, 0.910 + 29 = 29.910). At a frequency of 29MHz a single tuned circuit would give quite a wide bandwidth, and with the image response less than 1MHz away from the main response the attenuation of the image might well be only about 6dB or so! The situation could be improved by adding more tuned circuits, but it would take several in order to give really good results, and with each one needing a tuning gang this would be a little impractical.

The example given above is a rather extreme one, and at lower frequencies the situation would improve somewhat as the image response would stay 0.91MHz above the main response, but the lower operating frequency of the input tuned circuits would result in them having a narrower bandwidth. However, the fact still remains that a low IF and high reception frequency give poor image rejection, and this can make reception on the HF bands difficult using such a set.

One way of obtaining improved image rejection is to use a higher IF. An IF of say 9MHz would put the image response some 18MHz away from the main response, making it easy to obtain good image rejection even at the high frequency end of the short wave band. The problem with this set-up is that it is comparatively difficult to obtain high gain at a high IF, although this can be achieved without too much difficulty. Another drawback of a high IF is that ordinary L − C tuned circuits do not give very good selectivity at these frequencies, and this makes it essential to use a crystal filter (or some other type of high quality filter) in order to obtain really good performance. Provided the set is well designed and does incorporate such a filter, it will give extremely good results indeed.

Multiple Conversion

An alternative approach is to use a receiver having two or three frequency changer stages, and sets of this type are known as double and triple superhets respectively. The block diagram of Figure 12(a) shows the basic arrangement of a double superhet receiver.

The input signal is coupled to a tuned RF amplifier in the normal way, and then to a frequency changer stage which gives a first IF output at a fairly high frequency. The high first IF gives good image rejection, but makes it relatively difficult to obtain good gain and selectively. The output of the first frequency changer is therefore taken to a second frequency changer circuit which is not tunable. This one simply takes the first IF signal and takes it down to a much lower, second IF, where high gain and good selectivity can be achieved relatively easily. Thus a double superhet is able to obtain the best of both worlds, although at the expense of added complexity, and

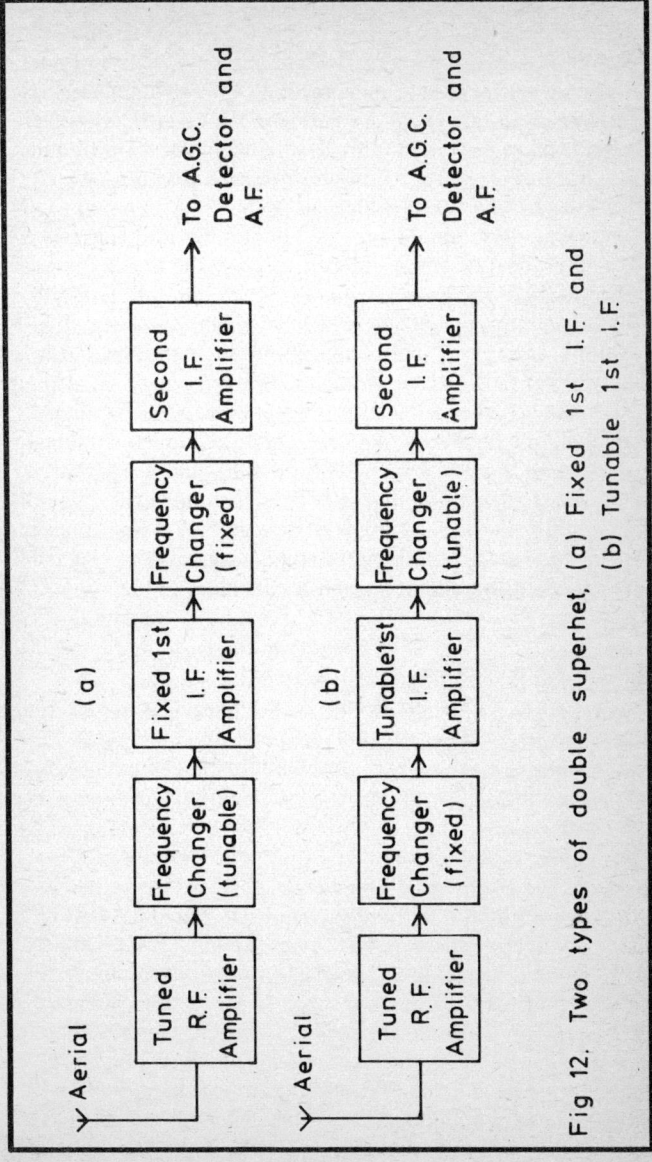

Fig 12. Two types of double superhet, (a) Fixed 1st I.F. and (b) Tunable 1st I.F.

31

practical designs of this type often have more spurious responses than a simple superhet, although these are usually not strong enough to be a serious problem.

A triple superhet is similar to a double type, but has an extra frequency changer stage which gives a really low final IF, typically around 50 to 100kHz, and sets of this type normally have very comprehensive, high performance IF filtering.

Although double and treble superhets of this type are regarded as rather old fashioned by many, these systems are still in current use, and sets of this type usually perform very well. Some sets work as single superhets on the LF bands, and double superhets on the HF bands. A similar system is often employed with triple superhets, and this seems to work well in practice.

An alternative form of double superhet is shown in Figure 12(b), and here the first frequency changer is not tunable, but the second one is. The first IF is not fixed, but is tunable over what is normally a fairly narrow range of frequencies (typically over a frequency span of about 1 or 2MHz).

If we take a simple example, let's suppose that we want to cover bands of frequencies 1MHz wide; this could be achieved by having a first IF which is tunable from (say) 4MHz to 5MHz. This range of frequencies is quite high, making it fairly easy to obtain good image rejection. If we wished to cover a tuning range of (say) 14 to 15MHz, then the RF amplifier would be tunable over this range (or it could be designed to give a fairly flat response over this range with increasing attenuation away from this band of frequencies). The frequency changer would have fixed tuning so that an input at 14MHz would give an output at 4MHz, and an input at 15MHz would give an output at 5MHz. In other words, the 4 to 5MHz tunable IF is made to tune over the required 14 to 15MHz tuning range by the addition of the first frequency changer.

The rest of the set is quite straight forward with the second IF being at a fairly low frequency where good gain and selectivity are achieved fairly easily. In fact this type of set could be regarded as a single superhet with a converter added at the input to give the required tuning range.

Wadley Loop

This is a system which is very popular in modern designs, and it has advantages of excellent image rejection due to the use of very high first IF (usually around 40 to 60MHz), and excellent tuning stability. In priniciple it is somewhat more complicated than most other types of superhet, as can be seen from the block diagram of Figure 13.

It should first be explained that the frequency changer stages we have discussed earlier are in fact comprised of two sub-stages: an oscillator and a mixer. The output from the mixer contains two frequencies that are generated by the mixing (heterodyne) action, and these are the sum and difference frequencies. For example, input frequencies of 2MHz and 10MHz would produce outputs at 12MHz (10 + 2 = 12, the sum frequency) and 8MHz (10 − 2 = 8, the difference frequency). If it is necessary for a superhet receiver to tune from (say) 10MHz to 20MHz with a 1MHz IF, the oscillator would be made to track 1MHz above the signal frequency, and would tune from 11 to 21MHz. The difference frequency would then provide the required 1MHz output. Alternatively, the oscillator could be made to tune a range 1MHz lower than the required tuning range, or 9 to 19MHz as that would be in this example, since this would also give a 1MHz output to the IF amplifier (in a simple conventional superhet it is, however, more common to have the oscillator above the signal frequency).

To take another simple example, if we wished to convert input signals in the range 20 to 21MHz to a tunable IF of 50 to 51MHz, an oscillator frequency of 30MHz would then be required, the sum signal giving output signals raised by 30MHz so that the required IF range is obtained.

As will be apparent by referring to Figure 13, a Wadley loop uses three mixers and two oscillators. Note that only the Wadley loop section of the receiver is shown in Figure 13, the rest of the receiver effectively just being a simple superhet which tunes over the appropriate IF output range. In this case the IF output is from 6 to 7MHz, but the figures shown in the diagram are only to help demonstrate the principle, and in practice will vary from one design to another.

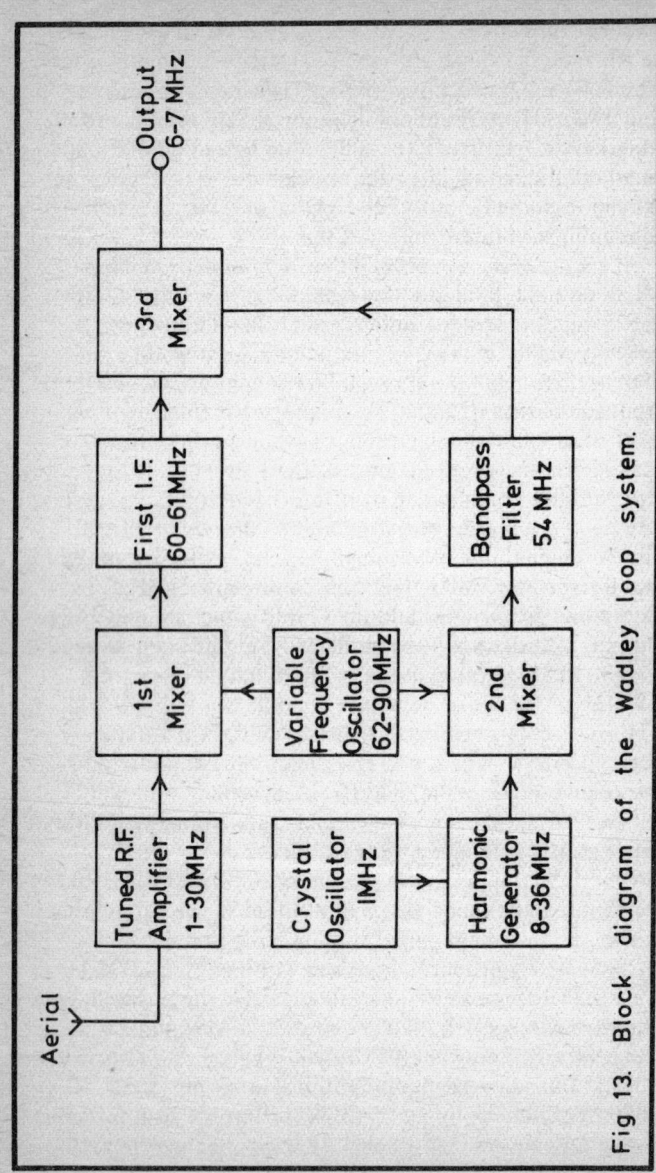

Fig 13. Block diagram of the Wadley loop system.

The set tunes from 1 to 30MHz in 29 x 1MHz wide bands. The RF amplifier therefore has 29 x 1MHz wide tuning ranges, or it can have 29 preset input filters having a bandwidth of about 1MHz. This would not be acceptable in most superhet arrangements, but remember that we are using a first IF of around 60MHz, giving an image response so far removed from the main response (about 120MHz) that there is no problem in obtaining high image rejection.

Let's assume that the set is required to tune from 29 to 30MHz, and that signals in this range have been selected by the RF amplifier and fed to the first mixer. The variable frequency oscillator would be tuned to 90MHz so that the difference frequency would give the required 60 to 61MHz output to the first IF amplifier. Like the RF amplifier, the first IF stage can either be made tunable over this range, or it can be designed to give good sensitivity over this entire range, with the response falling off outside the range.

Some of the output of the variable frequency oscillator (VFO) is taken to a second mixer. Here it is mixed with the output of a stable 1MHz crystal oscillator which has been processed to give strong harmonics. Harmonics are merely multiples of the fundamental frequency, and thus occur at 2, 3, 4, 5MHz, etc. In this case it is only the harmonics from 5 to 34MHz that are of interest.

The output of the second mixer is taken via a bandpass filter to one input of a third mixer. The bandpass filter only passes signals at or around 54MHz. A 54MHz output will be produced by the 90MHz VFO signal heterodyning with the 36MHz output of the harmonic generator. Many other output frequencies will be produced of course, and this is why it is necessary to have the bandpass filter to select the correct one of these.

The 54MHz signal is heterodyned with the 60 to 61MHz first IF signal at the third mixer stage to give the required output from 6 to 7MHz. The point of this system is that although a high frequency VFO is used (in order to give a very high first IF and good image rejection), excellent tuning stability is obtained due to a drift cancelling technique. If, for example, the VFO drifts slightly higher in frequency, this gives a higher difference frequency range at the output of the first mixer. However, the 54MHz difference frequency at the output of the second mixer also increases in frequency slightly, so that the difference frequency output from the 3rd mixer remains unaltered, and the tuning is totally unaffected by drift in the VFO.

Any drift in the 1MHz oscillator will result in a shift in the tuning, but as this is crystal controlled it is very stable indeed. Any drift in the tuning of the superhet circuit fed with the output signal will also affect the tuning, but obtaining good stability here is not too difficult since relatively low frequencies are involved.

Although on the face of it there is no way of altering the tuning using the VFO, this is not in fact the case. If its output frequency is reduced to 89MHz the receiver can tune from 28 to 29 MHz. A few quick calculations will show that this frequency range is required at the input to the first mixer in order to give the correct first IF range. The 89MHz VFO signal mixes with the 35MHz harmonic signal from the harmonic generator to give the required 54MHz output from the second mixer. In other respects the set operates as before. As will probably be apparent, reducing the VFO frequency in 1MHz steps reduces the frequency coverage of the set in 1MHz increments.

Receivers which use this principle are capable of excellent performance, and about the only drawback of this system is that it is obviously more complex than most. Sets of this type also need to be well designed in order to prevent problems with spurious responses, and to avoid problems with internally generated signals being picked up by the set. However, with proper screening and filtering where required these problems can be rendered insignificant, and should not be a serious drawback.

Choosing a Set

Having looked at the requirements for an amateur band receiver and the principles involved, how do you use this information in practice when choosing a receiver? As stated earlier, one essential is that some of the amateur bands should be covered by the receiver, and preferably all six bands should be available. With most transmissions on the amateur bands being SSB and CW types, a set having a CIO/BFO is also a necessity.

The list below gives some other important but non-essential features.

High first IF to give good image rejection.

Good filtering, preferably with switched bandwidths for SSB, and CW reception, and a low slope factor for each. Good tuning stability.

Strong mechanical construction and a backlash free tuning machanism.

High sensitivity and low noise level.

Good bandspread over the amateur bands.

Product detector and good AGC system.

RF gain control (or RF attenuator control).

Aerial trimmer.

Accurate frequency readout.

"S" meter.

Most of the new communications receivers available today seem to have all the features mentioned above, except for the comprehensive filtering. This is quite an expensive feature, and is something that is only likely to be found on one of the more expensive receivers.

All new receivers seem to be quite costly, but there are alternatives if such a set costs more than you are prepared to spend. Many amateur radio suppliers have stocks of second hand receivers, and many are offered for sale privately in the classified advertisement columns of amateur radio magazines. Many second hand sets use valve circuitry, whereas most (if not all) new sets are now fully solid state. Do not assume that this means valve circuitry is inferior and will give poor results: there are some classic sets of forty years ago that are still in use today and giving excellent results. On the other hand there are plenty of old (and not so old) receivers around that have deteriorated to the point where they are practically useless.

When buying a second hand receiver it is obviously advisable to actually try the set out and assess its condition before deciding whether or not its worth the asking price. This is not just a matter of checking the external appearance to see if the set has been well treated, make sure that all the controls are functioning properly, and check to see if the performance is what one would expect, checking all the bands covered by the set.

There are two problems one should bear in mind when buying one of the older communications receivers. One is simply that some of them are quite large and extremely heavy indeed, and a large strong bench is needed to accommodate such a set safely (if you ever try to pick up such a set you will see that I am quite serious about this)! The other is that it may be very difficult to get an old receiver serviced, and it might be better not to contemplate buying such a set unless you are capable of servicing the set yourself, or know of someone who would undertake the work for you. Even then, it is quite possible that suitable spare parts would not be obtainable.

An alternative to buying a new or second hand receiver is to build your own, although there are few people who could successfully build a complex communications receiver from scratch without first gaining some experience by constructing a simple receiver. Building ones own receiver usually gives a very considerable saving in cost over a ready-made alternative, and there is no reason why a home constructed receiver should not achieve a performance equal to that of commercial sets.

If you do start by building a simple short wave receiver you may well be surprised by the results that can be achieved. Most of the very simple short wave receiver designs for the home-constructor are of the TRF (tuned radio frequency) type, or the direct conversion type. A TRF set can usually be used for CW and SSB reception on the amateur bands, although this type of set is primarily intended for AM reception. With this type of circuit the signal is amplified at the received frequency only; no frequency changer or IF stages being used. In other words the RF amplifier feeds direct into the detector stage. A direct conversion receiver is much the same, but the detector is a product detector rather than an envelope (AM) type. Thus a direct conversion set is only suitable for CW and SSB reception, but for a simple amateur bands receiver this is perfectly acceptable.

It would be naive to expect a simple set of this type to out-perform a complex communications receiver, but I have picked up stations from all parts of the world using simple TRF and direct conversion receivers, and have had at least as much enjoyment from such sets as from more sophisticated sets. There is certainly a greater sense of achievement when a DX station is received using a simple set than when a station of similar difficulty is received using a complex receiver.

Some simple short wave receiver designs can be found in the book "Solid State Short Wave Receivers for Beginners", and some more sophisticated (superhet) designs can be found in the book "How To Build Advanced Short Wave Receivers", both by the same publisher and author as this book.

Whether you build your own receiver (simple or complex), buy second hand, or buy a new receiver then, must depend on how much you wish to spend or you have available for the purchase of a receiver, but whatever course you take you should derive many hours of pleasure from a very interesting and challenging hobby.

The Bands

Whatever receiver you finally obtain, in order to use it most effectively you need operating experience and a knowledge of the bands you are using. Obviously you can only obtain operating experience actually using the set, which presumably will not be too much of a problem! You can obtain a knowledge of the characteristics of each band by spending a lot of time listening on them, but you will obtain better results in the early stages if you are already primed with a knowledge of the bands and know what to expect.

Therefore we will next consider each of the bands in turn, giving consideration to the best times of search for DX stations, special difficulties involved, and things of this nature.

160 Metres

160 metres or "topband" as it is often called is a very difficult

band from the DXers pont of view. In Britain a maximum power of only 10 watts is permitted on this band (which compares to 150 watts on the other bands), and in many other countries there are similar restrictions. What is probably worse when trying to receive DX on this band is that it is shared with maritime transmissions, and the band is often quite crowded with powerful maritime transmission. Thus there is not just the problem of sufficient sensitivity to pick up the weak signals that result from the low powers used, there is the additional problem of filtering out strong interfering signals. Higher power operation on this band is allowed in some countries incidentally, but even these transmissions are likely to be much weaker than any comparitively local maritime or amateur transmissions, and by no means easy to copy properly.

During daylight hours the upper layers of the atmosphere allow signals at the frequencies involved here to pass through and into space. This severely limits the maximum range that can be obtained (usually not much more than about 100 miles or so) as the radio signals are not able to negotiate the curvature of the earth since they progate in a virtually rectilinear fashion. Except under exceptional conditions, daytime DX on topband is therefore impossible.

During the hours of darkness the situation is very different, with signals being reflected from a layer of the atmosphere known as the "E" layer. As can be seen from Figure 14, signals can be reflected back to earth at a considerable distance from the transmitter, giving a much greater range than is obtained from the direct path (known as the ground wave). The reflected signal is known as the sky wave and at the frequencies involved here provides good reception at ranges from a few hundred miles to a thousand miles or more.

Reception over greater distances requires the signal to be reflected from the atmosphere to the earth, from the earth to the atmosphere, and then back to earth again. Reception of really distant stations may require the transmitted signal to undergo a number of these reflections from the earth to the atmosphere and back to earth again. Each reflection causes the signal to be attenuated by a fairly substantial amount, but under favourable conditions it is possible for topband signals to be transmitted on one side of the world and successfully

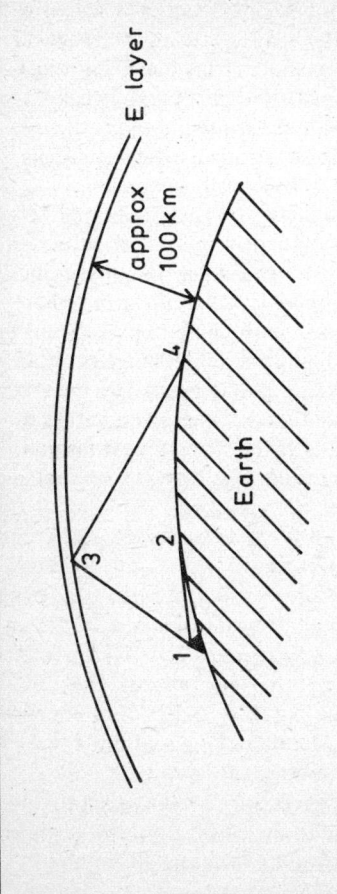

Fig 14. The ground wave (point 1 to point 2) has limited range as it cannot negotiate the curvature of the earth, but the sky wave (point 1 to point 3 to point 4) can, and gives far greater range.

received on the opposite side. However, this is not exactly an everday occurance, and could reasonably be regarded as the ultimate DX.

So, except under exceptional winter daytime conditions topband DX signals can only be received during the hours of darkness. Due to the difficulties involved in DXing on this band, CW is used a good deal by radio amateurs trying topband DX, and an ability to copy Morse code signals is a decided advantage. SSB is also used, but this is a slightly less effective transmission mode and does not offer quite the same DXing potential as CW. Nevertheless, SSB DX signals from European and North American stations can be received quite often, and reception over greater distances is sometimes possible.

Most of the topband DX I have heard has been well towards the low frequency limit of the band, and has been either late at night or in the early hours of the morning. Winter-time seems to produce more DX signals than does summertime. DX stations will not often be found on a completely unoccupied channel, and are usually in very close proximity to a more local transmission of some kind. A receiver having really good selectivity is therefore a considerable asset for this type of reception. However, the DX station and the interfering local station will often overlap to some extent, and no matter how good the selectivity of the receiver, the DX station will have to be copied through a significant amount of interference. The types of station (non-amateur ones that is) are such that the interference will probably be continuous, and hard concentration will be required in order to properly copy DX stations.

When trying to copy a station through a lot of noise or interference it is usually somewhat easier using headphones rather than listening via a loudspeaker, and I would strongly recommend the use of headphones for any form of DXing. Special headphones for use with communications receivers are available, and these are designed to have a frequency response that is flat over the range of frequencies needed for good voice communications, but rolls off rapidly outside these limits. This helps to minimise noise and interference, although a good receiver will give very little audio output at frequencies outside these limits, and communications headphones will be of comparatively little benefit with such a receiver. However, if you are buying headphones for communications purposes it is obviously better to have the proper type than to use hi-fi headphones (which are designed to have as flat a response as possible over the entire audio frequency range).

A problem that is often encountered when using headphones is that of noise spikes received by the set causes ear splitting "clicks" and "bangs" which can verge on the hearing threshold of pain! Even if the noise spikes produced are of tolerable volume, heavy interference of this type can make it very difficult to copy even quite strong stations.

The ideal way of counteracting this type of interference is to use a device called a "noise blanker". In effect, this cuts off the audio signal very briefly when a noise spike is detected, causing the loud noise pulse to be replaced with a gap in the signal. As the gap will be too brief to be heard as such, and is normally barely audible or not audible at all, this can give a very substantial increase in the intelligibility of a signal. Unfortunately, few receivers incorporate a noise blanker, and it is a feature that is normally only found in professional communications systems and the most expensive of amateur communications equipment. It is not an easy matter to add one to an existing receiver either.

An alternative is to use a device known as a "noise limiter", and this is a feature of some amateur receivers, although it is not used as much as one might expect for such a simple but useful type of circuit.

It is quite simple to add a noise limiter between the audio output of the receiver and the headphones, a suitable design being shown in Figure 15. This is extremely simple and costs very little to build. Also, it is a passive circuit and does not require a battery or other form of power supply.

D1 and D2 are silicon rectifiers, and require a forward bias of about 0.5 volts in order to bring them into a state of conduction. Raising the forward bias only fractionally causes the device to avalanche and a high current to flow. In other words the rectifier acts rather like a zener diode having a zener voltage of only about 0.5 volts.

This circuit is really just a type of shunt stabiliser, with R1

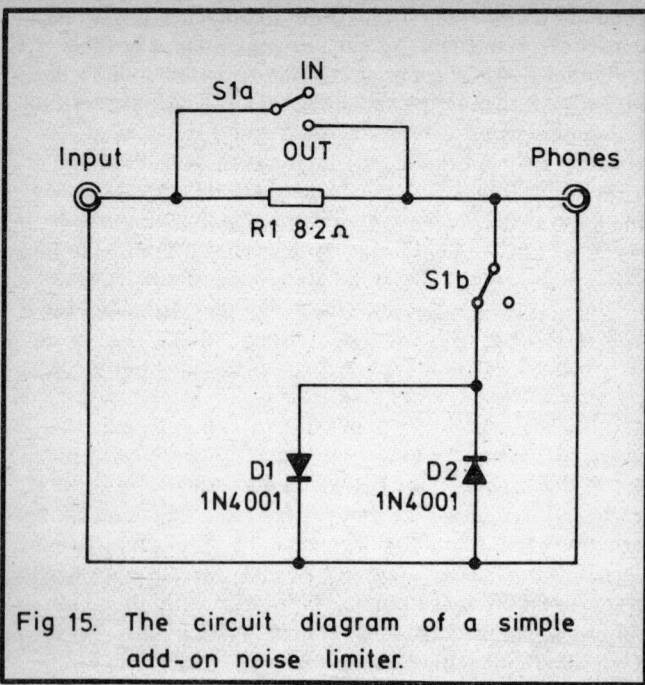

Fig 15. The circuit diagram of a simple add-on noise limiter.

acting as the load resistor, D1 limiting the positive output potential to no more than about 0.5 volts, and D2 limiting the negative output voltage to a similar level. Provided the voltage across D1 and D2 does not reach 0.5 volts the circuit has no effect, except a small attenuation of the signal due to the voltage drop through R1. This can be counteracted by a small advance in the volume control (or AF gain control as it is often called in this application).

If signal peaks should try to exceed the 0.5 volt threshold level of the rectifiers, they will fail to do so because the rectifiers will conduct on the peaks, causing an increased voltage drop through R1 and preventing the peaks from rising significantly above the 0.5 volt threshold level.

In use the volume control is advanced so that normal signals are just below the clipping level, or are just slightly clipped by the action of the circuit. In practice this is just a matter of advancing the volume control as far as possible without serious distortion occuring. Any noise spikes will be limited to a peak level that is no more than the peak level of the wanted signal. This prevents the noise spikes from reaching deafening levels, and increases the signal to noise ratio when there is heavy interference of this type because the noise spikes will be seriously attenuated, whereas the wanted signal is virtually unaffected. This effect is illustrated in Figure 16.

S1 is merely a bypass switch that enables the limiter to be switched out of circuit if desired.

Components: Noise Limiter, Figure 15

Resistor
R1   8.2 ohms ¼ watt 5%
Diodes
D1   1N4001   D2   1N4001
Switch
S1   DPDT toggle type
Miscellaneous
Case, input and output sockets, wire, etc.

Notch Filter

A common form of interference on the low frequency bands, and one that can be very troublesome when topband DXing, is that of heterodyne tones. These are caused by carrier waves reacting with the BFO or CIO signal to produce a continuous audio output tone.

This type of interference can be eliminated using a notch filter operating in either the IF or audio stages of the receiver. A few receivers have a built in notch filter, but it is not a common feature of communications receivers. However, it is quite easy to add an audio notch filter at the audio output of a receiver, and the improvement in intelligibility that can be obtained by using such a filter (when a strong heterodyne is present) is very considerable.

The circuit diagram of an add-on audio notch filter is shown

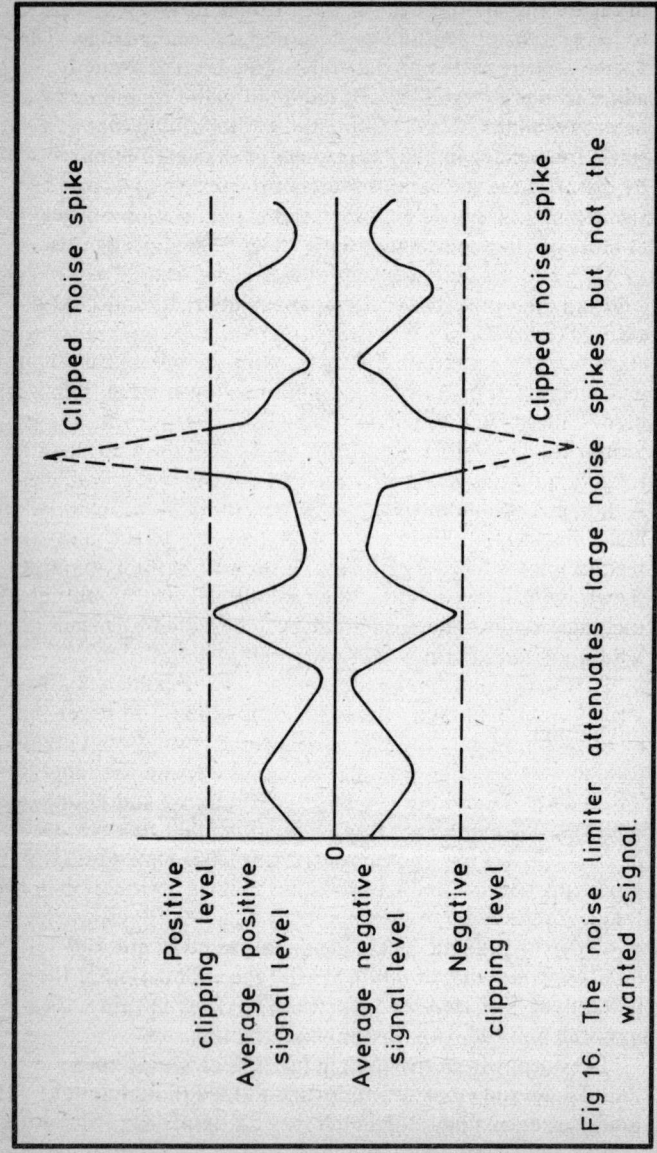

Fig 16. The noise limiter attenuates large noise spikes but not the wanted signal.

46

in Figure 17, and this can be tuned from a little under 100Hz to beyond the upper limit of the audio frequency range. The degree of attenuation provided when the filter is properly adjusted is in excess of 80dB, and is sufficient to render any heterodyne inaudible. Of course, the filter only works at a single frequency, and any harmonics of the heterodyne caused by distortion in the various stages of the receiver will not be significantly affected. However, this should not be a problem as most receivers have only low levels of harmonic distortion.

The circuit is quite conventional, and Tr1 is used as a phase splitter with out-of-phase signals appearing at its collector and emitter terminals. In effect, Tr1 operates as a common emitter stage from its base to its collector, giving an inverted signal at its collector. It operates as an emitter follower stage from its base to its emitter, giving no phase change between these two terminals. The voltage gain from the base to the emitter is roughly unity, as is normally the case for an emitter follower. Although a common emitter amplifier normally has a fairly high voltage gain, this is not the case here due to the negative feedback introduced by R3, and this results in the two output signals being of roughly the same amplitude. The amplitude of the signal at the collector terminal can be varied by means of VR1.

The two outputs are fed to a Wien network which consists of VR2, C2 and C3. At a certain frequency there will be an identical phase shift through each section of the Wien network (one section being formed by VR2a plus C3, and the other by VR2b and C2), and the two outputs will have a cancelling effect on one another. VR1 is adjusted so that the two signals precisely cancel one another out at this frequency, thus producing the required narrow notch of high attenuation in the frequency response of the circuit.

VR2 is the tuning control, and is adjusted to null the interfering heterodyne. To optimise the attenuation of the heterodyne VR1 and VR2 should be adjusted in turn until it has been reduced to an insignificant level.

The output from the Wien network is at a fairly high impedance, and a buffer amplifier is needed to match this to a loudspeaker or a pair of headphones. A simple amplifier using an LM380N IC (IC1) is used to provide this buffering. The

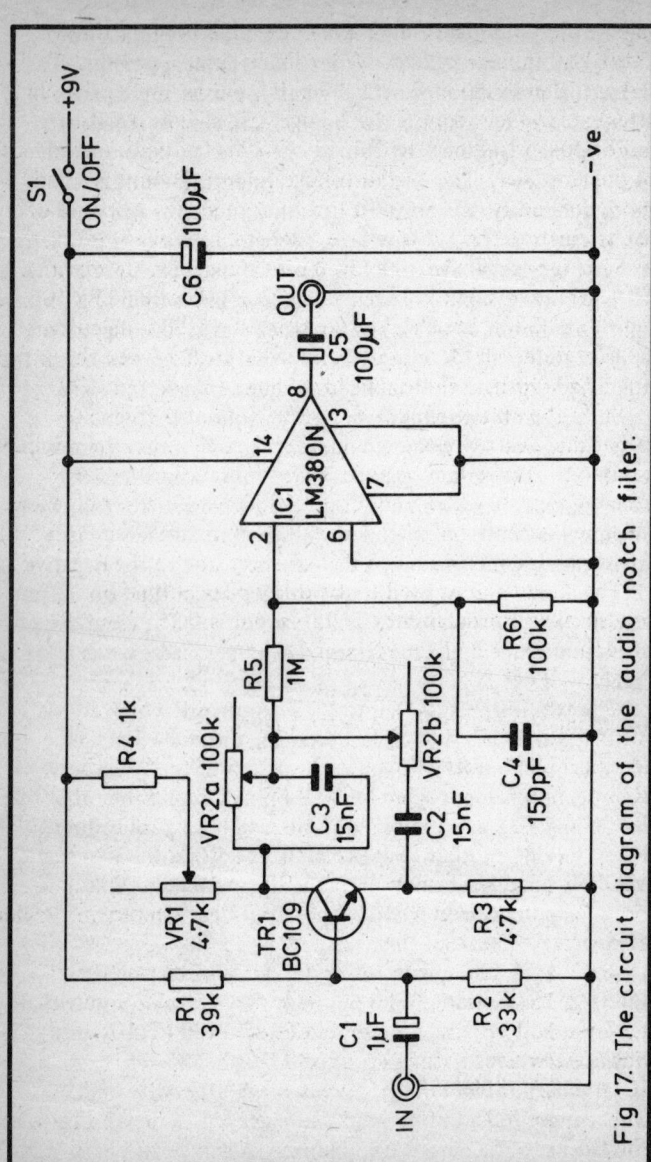

Fig 17. The circuit diagram of the audio notch filter.

48

gain of the LM380N is fixed at a nominal level of 50 times (34dB) by an internal negative feedback circuit, and this is rather higher than is required here. R5 and R6 are used as an attenuator which reduces the voltage gain of the circuit to a more appropriate level, and R5 also boosts the input impedance of the amplifier. The overall voltage gain of the unit is a little more than unity. C4 rolls off the high frequency response of the circuit and this helps to give good stability.

S1 is the on/off switch and C6 provides supply decoupling. The current consumption of the circuit is approximately 10mA, but the LM380N has a class AB output stage, and the current consumption will therefore be somewhat higher when the unit is driving a loudspeaker at fairly high volume.

Since the unit attenuates a narrow band of frequencies rather than just affecting one frequency, the unit can sometimes be used to good effect on types of interference other than heterodynes. In order to be effective though, the interference must be concentrated mainly over a fairly narrow band of frequencies (which does quite often seem to be the case).

The circuit can be used to drive any normal type of loudspeaker or headphones, and its input is taken from the headphone socket of the receiver.

Components: Audio Notch Filter, Figure 17

Resistors, all ¼ watt 5%

R1	39k	R2	33k
R3	4.7k	R4	1k
R5	1M	R6	100k
VR1	4.7k lin. carbon potentiometer	VR2	100k + 100k lin. carbon potentiometer

Capacitors

C1	1 μF 10v	C2	15nF plastic foil
C3	15nF plastic foil	C4	150pF ceramic plate
C5	100 μF 10v	C6	100 μF 10v

Semiconductors

IC1	LM380N
Tr1	BC109

Switch

S1	SPST toggle type

Miscellaneous

Case, component panel 9 volt battery, input and output sockets, wire, solder, etc.

80 Metres

In a way conditions on 80 metres are much the same as on topband, with short range reception during the day and longer distances being covered at night. Also like topband, it is not band space that has been set aside exclusively for amateur use, and a certain amount of interference from commercial stations has to be tolerated.

There are important differences between the two bands though, and whereas topband is used mainly as a band for local contacts between stations often just a few miles apart, much larger distances are usually involved on 80 metres, even during daylight hours. In fact it is possible to receive Continental amateur stations during the daytime on this band in most parts of the country. After dark it is not uncommon to pick up amateur stations a few thousand miles away, and worldwide reception can be achieved much more easily and frequently than on topband.

There are several reasons for the better reception on 80 metres, and one is simply that there seems to be less interference from commercial stations. Another is that higher power levels are generally employed, and this band is also more popular. At the higher frequencies involved on this band signals also seem to be propogated more efficiently, giving increased range for what are in other ways comparable operating conditions.

Like topband, in general the best time for DXing is late at night and in the small hours of the morning. Remember that in some countries the band extends up to 4MHz (not 3.8MHz as in the U.K.), and that it is often well worth exploring the additional 3.8MHz to 4.0MHz segment of the band. Quite often North American stations can be heard here fairly late at night.

40 Metres

This is a rather difficult band and at one time was not at all popular. It seems to be used more these days, but is still not the most popular of bands. In theory it should have excellent DX properties, behaving in a similar manner to 80 metres, but with signals propogating better both during the daytime and at night. This is in fact the case, and many strong continental amateur transmissions can usually be received during the day, and stations from South America and further afield can often be received at good strength after dark.

The problem with this band is firstly that it is only 100kHz wide, and it is positioned right next to the 41 metre (7.1 to 7.3MHz) broadcast band. The small amount of band space obviously limits the amount of stations that can be accommodate at one time, and although in some countries the band extends up to 7.3MHz, the additional part of the band coincides with the 41 metre broadcast band. In Europe this broadcast band is much used by very powerful stations that make it very difficult to pick up the relatively weak amateur transmissions on this band.

Unfortunately, these stations are also apt to encroach onto the supposedly exclusive 7.0MHz to 7.1MHz part of the band, and these can cause problems with interference, especially after dark when they are usually stronger and more numerous.

Apart from any in-band broadcast stations, broadcast stations in the 7.1MHz to 7.3MHz segment can also give rise to problems. Although such stations will lie outside the IF passband of the receiver, they will almost certainly be well within the RF passband of the set. As pointed out earlier, strong signals in the RF stages of a receiver can cause overloading here, with consequent distortion and cross modulation. At night the stations on the 41 metre broadcast band are so strong and numerous that even with a very well designed set, cross modulation is almost inevitable if the set is operated with the RF gain control (or input attenuator) well advanced.

It is therefore usually necessary to keep the RF gain control well backed off when DXing on the 40 metre band after dark. This will effectively reduce the strength of received signals, but will probably give a much larger decrease in the cross modulation noise.

Bandpass Filter

A useful improvement in performance on 40 metres can

often be obtained by adding a simple bandpass filter between the aerial and the receiver. This increases the RF selectivity, so that the wanted amateur transmissions are made stronger in comparison to the interfering broadcast stations (although as the unit is passive, all signals are reduced in strength by its addition, the amateur signals being attenuated less than broadcast stations).

The circuit, as shown in Figure 18, just consists of a tuned circuit formed by the main winding of T1 and VC1, with a coupling coil on T1 being used to introduce the aerial signal into the circuit. C1 loosely couples the tuned circuit to the input of the receiver. A stronger output signal can be obtained by increasing the value of C1 slightly, but this would tend to flatten the response of the filter somewhat and give reduced performance. Reducing the value of C1 will give a weaker output signal, but the response of the filter would be improved

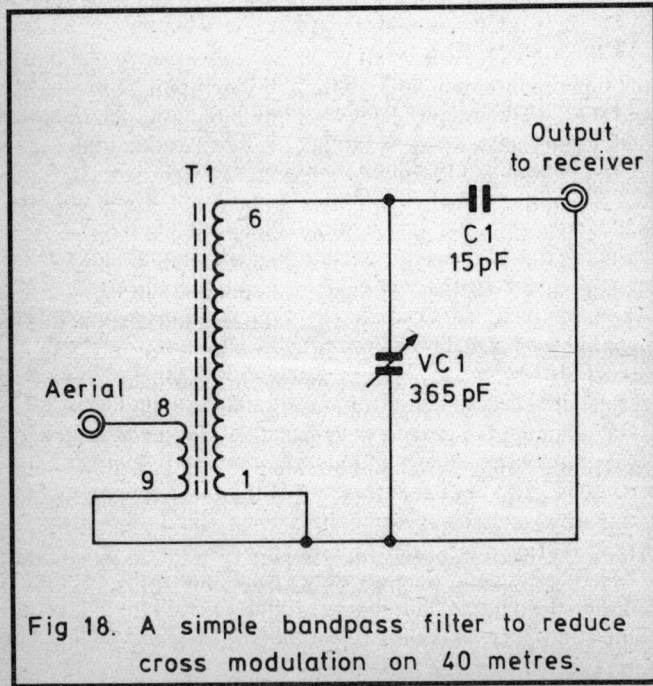

Fig 18. A simple bandpass filter to reduce cross modulation on 40 metres.

slightly. The specified value is the one that seems to give the best overall performance.

The output cable to the receiver should be as short as possible, and should preferably be a coaxial type (the outer braiding taking the earthy output of the filter to the earth connection of the receiver, and the inner conductor taking the "hot" output of the filter to the aerial terminal of the receiver). A non-coaxial cable can be used, but signal pick up in this would effectively reduce the performance of the filter, although perhaps not seriously.

Incidentally, the added RF selectivity provided by the filter will give improved image rejection as well as reduced cross modulation.

Components: Bandpass Filter, Figure 18

Capacitors
C1   15pF plastic foil
VC1   365pF air spaced (Jackson)
Inductor
T1   Denco DP aerial (blue) coil, range 4
Miscellaneous
Case, sockets, control knob, wire, solder, etc.

20 Metres

The three low frequency amateur bands have similar propogation characteristics, but things become very different on the three high frequency bands. On the low frequency bands the ground wave signal provides communication over relatively short distances at any time of day, and at night the skywave signal provides communication over much greater distances. On 20 metres the ground wave signal tends to be absorbed by the earth, giving range of only a few miles. In fact for most practical purposes the absorption of the ground waves makes this signal unusable. During the hours of darkness the sky wave is often nonexistent, and the band may well go completely "dead" after dark. The sky wave signal is normally strong during the daytime, but it is not reflected from the E layer of the atmosphere as is the case on the low frequency bands, but is reflected from the F2 layer.

As shown in Figure 19, by reflecting from a higher layer of the atmosphere the signal achieves greater range per reflection. Line A – B – C represents the path of a low frequency signal reflected from the E layer, and line A – D – E represents the path of a high frequency signal reflected from the F2 layer. Line A – F represents a high frequency signal radiated at a high angle, and this passes through the F2 layer and on into space. Thus there is no short range communications via the ground wave, and signals reflected from the F2 layer of the atmosphere are not able to provide short range communications either.

This absence of strong local signals and long range per reflection makes 20 metres and the other two high frequency amateur bands ideal for long distance reception. Even using a quite simple receiver, it is not that difficult to receive stations from all parts of the world.

Whereas DX reception on the low frequency band requires a path of darkness between the transmitter and receiver, DX reception on the high frequency bands require a path of daylight. Thus on the low frequency bands the longest distance reception occurs to the east as darkness falls, and to the west just before dawn. On the high frequency bands pretty much the opposite occurs, with maximum range to the east occuring just after dawn, and optimum range to the west occuring just before dusk. Reception of stations on the other side of the world when listening on 20 metres at these times is not a rarity, with the early mornings tending to give better results in practice than the evenings.

Long distance reception on 20 metres is so common-place that many do not consider a station on this band to be DX unless it is both distant and rare (such as a station on a small group os islands where there are only a few amateurs).

In the middle of summer when the days are long and the nights are quite short, 20 metres often provides reception for 24 hours per day, albeit that the number of stations and their strengths are both greatly reduced during the middle of the night.

During the rest of the year it is normal for the band to become completely devoid of stations soon after dark, but at times of high sunspot activity the band may still be usable for

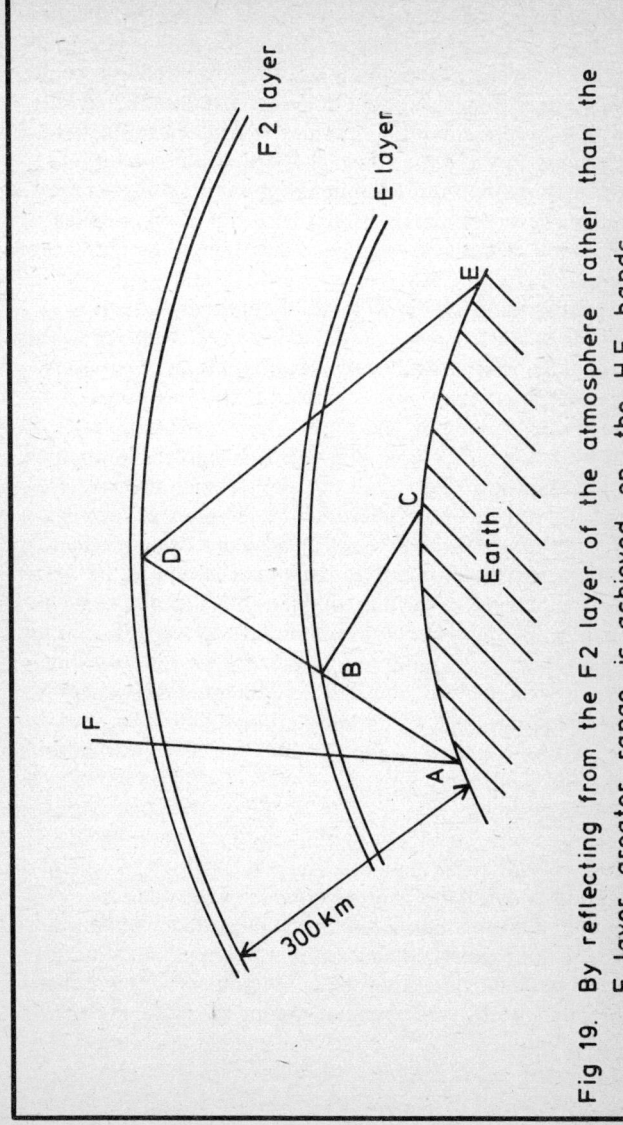

Fig 19. By reflecting from the F2 layer of the atmosphere rather than the E layer, greater range is achieved on the H.F. bands.

some hours after darkness. Sunspots are cool (relatively speaking) spots on the surface of the sun, and are accompanied by the release of vast amounts of energy from the sun's surface, some of this energy being received by the earth's atmosphere. Here it has an ionising effect on certain layers, making them reflective to high frequency radio waves, and this effect can persist well after sunset.

Sunspot activity is not consistent, and tends to run in roughly eleven year cycles with large numbers of sunspots appearing at peaks, and long periods with none at all during sunspot minimums. At the time of writing a sunspot peak has just been passed, and the next is not expected until about 1990. Best DX reception on the low frequency bands incidentally, is associated with sunspot minimums.

One last but important point about 20 metres is that due to its high DX potential it is probably the most popular of the bands. This has its good side and its bad aspect. On the plus side there is the fact that you can pick up stations from practically anywhere in the world, since anywhere there are radio amateurs there is likely to be someone using this band. Of course, stations a relatively short distance away cannot normally be received on this band, but these areas can be covered well by the low frequency bands. On the deficit side, the band is normally so overcrowded that it is necessary to copy DX stations through heavy interference from numerous stations on nearby frequencies. A receiver having really good selectivity is helpful here, but much of the interference is often within the same band space as the wanted signal, and there is no alternative to putting up with it.

15 Metres

This band is very similar to 20 metres in many ways, but it is not quite as popular, and at times of low sunspot activity it tends not to give quite such good reception as 20 metres, and dies out more quickly at the onset of darkness. Many people believe that this band is unusable except when there is a reasonably high level of sunspot activity, but in my experience this is not the case. Even with no sunspot acitvity it seems to be possible to receive some stations that are many thousands of miles away, but a good receiver and aerial are required. There are fewer stations than can be found on 20 metres, but this could just be due to lack of use rather than propogation conditions.

When sunspot activity is high, 15 metres has a DX potential that is at least equal to 20 metres, and is often superior. After dark though, 15 metres tends to die away earlier and more rapidly than 20 metres, and it is comparitively rare for this band to provide reception for 24 hours per day.

10 Metres

This band is very different to the other two high frequency bands in certain respects. One relatively minor difference is that there is less absorption of the ground wave signal by the earth, and local communications is possible, although only a quite small amount of local communications takes place on this band. Most local communications seems to take place on the 2 metre VHF amateur band these days.

During periods of low sunspot activity the 10 metre band is completely free from DX signals for the majority of the time. When it does pick up occasionally, signal strengths are generally quite low, and without a good aerial and sensitive receiver you may still receive nothing. However, communications over distances of several thousand miles can be achieved during these lifts in conditions.

When sunspot activity is low there is no DX communications on this band once darkness has fallen.

At times of high sunspot activity the 10 metre band is a totally different proposition. Results obtained will vary considerably according to the exact conditions prevailing at the time, but frequently good DX reception is possible. When conditions are really good this is certainly the best DX band. Worldwide reception becomes possible using low power transmitters and relatively simple receiving equipment. Stations on the opposite side of the world can be received very strongly indeed.

About the only problem at these times is that a great many radio amateurs try to take advantage of the favourable conditions while they have the chance, and it can sound like 20 metres at a time of peak use, or even worse. A receiver having good selectivity is an advantage, and a directional aerial is also of considerable help. Although many radio amateurs use directional aerials on this band, few short wave listeners seem to bother despite the advantages. A short beam aerial for this band does not need to be physically all that large, and has the property of giving increased signal strengths from signals emanating from whatever direction the aerial is aimed, and reduced signal strength away from this direction. Thus by setting up the aerial so that it is aimed in the direction of the required DX, these DX signals are boosted while most other stations will be attenuated (although a directional aerial cannot be guaranteed to give such an improvement in every case, as it will occasionally happen that the DX signal and an interfering signal are both coming from virtually the same direction).

Unfortunately, even at times of high sunspot activity the 10 metre band tends to die out quite quickly after dark, and it is not really suitable for night time DXing.

Preselector

Many receivers do not work especially well on the high frequency bands, the two main problems being a fall off in sensitivity, and rather poor image rejection if the set has a fairly low IF. To some extent these problems can be overcome by using a preselector ahead of the receiver. This is a tuned RF amplifier which gives both a boost in gain, and reduces the RF bandwidth of the receiving equipment. Thus both sensitivity and image rejection are improved by the addition of a preselector.

The preselector circuit shown in Figure 20 is tunable from about 10MHz to over 30MHz, and is therefore suitable for use on all three high frequency amateur bands. It provides a voltage gain of about 20dB (10 times) and has a low noise level.

A coupling winding on T1 is used to inject the aerial signal into the tuned circuit which is formed by VC1 and the main winding of T1. VC1 is, of course, the tuning control, and in practice is adjusted to peak received signals, taking care to boost signals on the main response. If in doubt as to which is which, the main response is normally the one which is peaked with VC1 at the higher capacitance setting (with its vanes

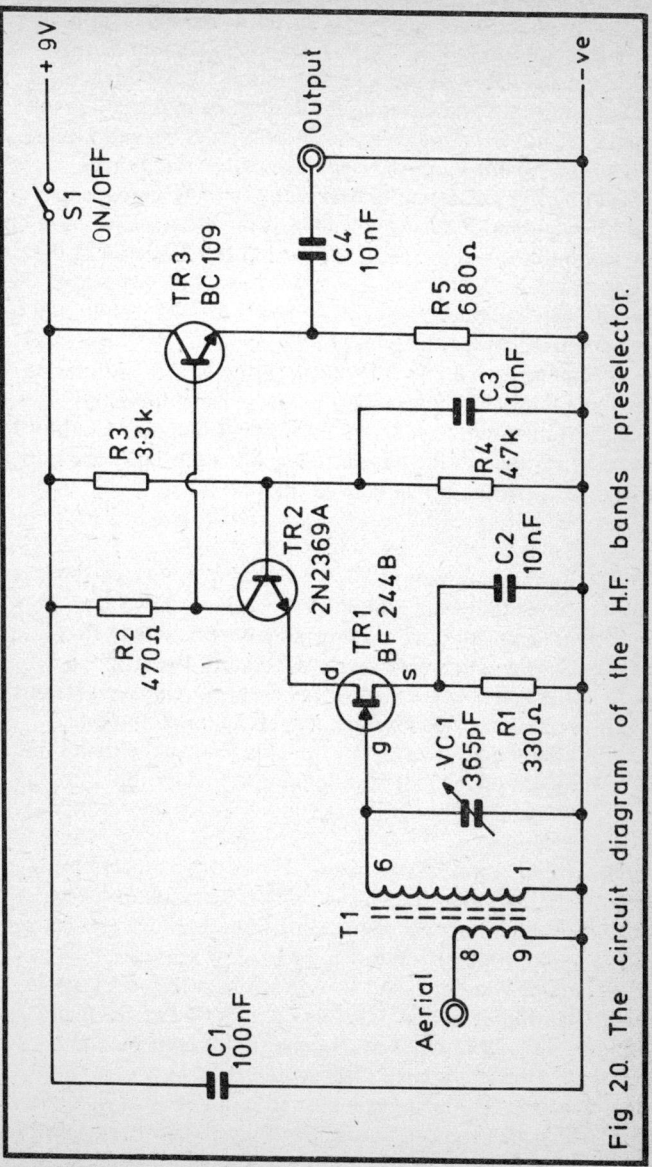

Fig 20. The circuit diagram of the H.F. bands preselector.

59

more fully meshed). Tr1 is a JFET device which has an extremely high input impedance, and the tuned circuit can therefore be coupled direct into its gate terminal.

Tr1 is used in the common source mode, and has R1 and C2 as its source bias resistor and bypass capacitor respectively. Its output (drain terminal) is fed direct to the input of Tr2 which is used as a common base amplifier. This is biased by R3 and R4, and has R2 as its collector load. Tr3 is an emitter follower buffer stage which is used to match the relatively high output impedance of Tr2 to the low input impedance that is presented by the aerial inputs of most receivers. R5 is the emitter load resistor for Tr3 and C4 provides DC blocking at the output. S1 is a straight forward on/off switch and C1 is the only supply decoupling component that is needed. The current consumption of the circuit is approximately 10mA.

The layout of the unit and method of construction used are not highly critical, but as with any high frequency circuit, long wires should be avoided as they could either produce a loss of performance or instability.

Use a cable that is as short as possible to couple the output of the preselector to the input of the receiver. Ordinary insulated leads could be used to carry the earthy and "hot" outputs of the unit to the receiver, but signal pick-up in the "hot" lead could significantly degrade the amount of extra RF selectivity provided by the unit. It is therefore better to use a short coaxial cable.

Components: HF Bands Preselector, Figure 20

Resistors, all ¼ watt 5%

R1	330 ohms	R2	470 ohms
R3	3.3k	R4	4.7k
R5	680 ohms		

Capacitors

C1	100nF ceramic	C2	10nF ceramic
C3	10nF ceramic	C4	10nF ceramic
VC1	365pF air spaced (Jackson)		

Semiconductors

Tr1	BF244B	Tr2	2N2369A
Tr3	BC109		

Switch
S1   SPST toggle type
Inductor
T1   Denco DP aerial (blue) coil, range 5
Miscellaneous
Case, sockets, 9 volt battery, control knob, B9A valve-holder (coil-holder), wire, solder, etc.

Call-Signs

Amateur radio stations all have a unique identifying call-sign, and this has a prefix which denotes the country from which the amateur is operating. Most amateur call-signs consist of one or two letters followed by a single number, and then two or three more letters. It is the first one or two letters that indicate the country in which the station is being operated, and the number sometimes indicates the particular part of the country concerned. However, in most cases the number is of little or no significance.

A map of the world giving reasonably up to date information on amateur radio prefixes is available from the Radio Society Of Great Britain at a modest cost, and is well worthwhile having. The RSGB is the national amateur radio club of the U.K., and membership is open to anyone who can show a keen interest in amateur radio. Members receive the interesting and useful monthly club magazine "Radio Communication". Most amateur radio clubs in Britain are affiliated to the RSGB, and they should be able to put you in touch with your local club. The address of the RSGB is 35 Doughty Street, London WC1N 2AE.

Returning to the subject of call-signs, the country from which the amateur is operating can often be found out without knowing what country is signified by the call-sign, as he or she will often give the location quite accurately, enabling the position of the station to be estimated to within a few miles. Since call-sign prefixes are changed from time to time, and new ones seem to be continually coming along, this is probably just as well as it would otherwise be difficult to keep up to date.

Phonetics

It is common for amateur radio call-signs to be given using a phonetic alphabet in order to reduce the risk of the call-sign being incorrectly copied. For example, the call-sign Z5ABC might be given as "Zulu Figure 5 Alfa Bravo Charlie". I would give the phonetic alphabet, but there are actually several in use, and many radio amateurs seem to invent their own! This lack of standardisation partially negates the advantage of using phonetics, and can lead to confusion, although it is generally easier to copy a call-sign given this way than one given using straight forward letters.

Q Codes

Q codes are a number of codes which consist of the letter "Q" followed by two other letters. These originated in the days when messages were only sent in Morse code, and the use of these three letter codes could greatly speed things up. These days Q codes tend to be used to some extent by amateurs when using telephony (voice communication) as well as when using telegraphy (Morse code communication). There is an extensive list of Q codes which have very precise meanings, but only a few of these are used to any great extent in amateur radio, and their meanings have become somewhat modified over the years. They do not, therefore, necessarily have the same meaning in the realm of amateur radio as in other fields of radio communication.

The list given below shows the Q codes in common use by radio amateurs and the meaning of each code when used by radio amateurs. In some cases the meanings are totally different to those of the same codes when used in professional radio circles!

QRM	Man made interference (adjacent channel interference, etc.)
QRN	Atmospheric interference
QRO	High power (transmitter)
QRP	Low power (transmitter)
QRT	Signifies that the station is closing down
QRX	Please wait

QRZ	You are being called by . . .
QSB	Your signals are fading
QSL	Did you copy that? Also a type of card sent as confirmation of a contact (or reception of a station)
QSO	A contact with another station
QSY	I am changing frequency
QTH	Location of the station.

VHF Bands

There are two VHF amateur bands in the U.K. (but not most other countries):—

4 Metres	70.025MHz to 70.700MHz
2 Metres	144.0MHz to 146.0MHz

The 4 metre band has little in the way of DX prospects since it is not an international band. In most countries there is either no amateur band towards the low frequency end of the VHF band, or a different frequency allocation is used (such as the American 6 metre amateur band). This has probably contributed to the lack of popularity for this band.

These days the 2 metre band is much used for both local and DX operating, and must rate as one of the most popular bands. However, DXing is a relative term, and results obtained on 2 metres do not compare with those on (say) the 20 metre short wave band. The ground wave travels quite well at these frequencies and the band is very suitable for contact over distances of up to about 50 miles or so. The range obtained obviously depends to a large extent on the equipment used, but the local terrain also has a significant effect. Hills, mountains and even large buildings tend to reflect or absorb VHF signals, and one would therefore normally expect to obtain better local reception when operating from a hill top than when operating from the bottom of a valley. In fact local reception is practically non-existent at some unfavourable locations!

DX reception is somewhat less affected by local conditions as the radio waves are reflected downwards from the atmosphere so that they are able to negotiate quite large obstructions. The problem is that at these very high frequencies it is uncommon for conditions to be favourable for DX reception, and signals normally travel straight through the atmosphere and into space. Even under favourable conditions the range attained at these frequencies is usually no more than a few hundred miles, although reception over greater distances is not unknown.

There are two types of atmospheric disturbance which permit DX on the VHF bands; Sporadic E and Tropospheric disturbances. The former usually occurs during April to August, and the patches of highly ionised gas which reflect the radio signals usually only last a few hours. It is because of their sporadic nature and the fact that they occur in the E layer of the atmosphere that the name Sporadic E is derived.

Sporadic E can provide reception over distances in excess of one thousand miles, but unfortunately they usually only affect signals from about 30MHz to 80MHz. It is therefore only under exceptional circumstances that this type of reception is possible on the 2 metre band. It is of little use on the 4 metre band due to the lack of amateur stations outside the U.K. using this band, and it often results in 4 metres being blotted out by European FM broascast stations.

Tropospheric reception is usually associated with high pressure areas (above 30in.) and can provide reception over distances of a few hundred miles or more. It affects frequencies from about 100MHz up to several hundred MHz, and is the form of DX reception most often encountered on the 2 metre band.

Equipment

If you already own a good short wave receiver, probably the least expensive way of obtaining good results on the 2 metre band is to use a 2 metre converter in conjunction with the short wave set. Such a converter takes signals in the 144 to 146MHz range and converts them down to a 2MHz wide band in the short wave spectrum. The most popular output range these days seems to be 28 to 30MHz (which is covered by most general coverage and amateur-bands-only receivers). Thus, with the converter feeding into the receiver's aerial terminal, by tuning the set from 28 to 30MHz (or whatever output range the converter happens to have) the set is effectively tuned from 144 to 146MHz.

One problem with this simple arrangement is that the mode of transmission which seems to be most popular on the 2 metre band at present is narrow band frequency modulation (nbfm), and there are few short wave receivers which are equipped to resolve such signals. This is not to say that nbfm cannot be resolved on the average short wave set, as it often is possible to do so.

The carrier wave of the received station must be tuned so that it is out of the passband of the receiver, but only just, as shown in Figure 21. It does not matter which side of the passband it is tuned to unless one side gives less adjacent channel interference than the other. As the carrier is modulated up and down in frequency, the output from the receiver's detector stage increases as the carrier moves towards the passband of the receiver, and decreases as it moves away. This varying signal is the demodulated audio signal.

There are two main problems with this solution. One is simply that with the wanted signal not in the passband of the receiver, some other signal might well be! As the VHF bands are not, as yet, especially overcrowded, in practice this does not often seem to be a great problem. Of more importance is the fact that the edges of the receiver's passband, even at the widest bandwidth setting, may be too narrow to accommodate the signal. This can result in the carrier being virtually switched from full strength to practically nothing as it varies in frequency, giving a totally distorted and unusable audio output. Thus a set with a relatively wide bandwidth and poor slope factor is likely to give better results in this application than one having very high performance filtering.

SSB is used to some extent on the 2 metre band, especially by amateurs who are trying to make DX contacts. Obviously there should be no problems in resolving this type of signal, and I have obtained good results on SSB using a 2 metre converter feeding into a short wave set. CW is also used, and the same is true of this mode. Obviously there should be no problem in resolving ordinary AM signals, but this mode is now a rarity on the 2 metre band.

Mainly DXing takes place in the 144 to 145MHz section of the band, the other half being used principally for local contacts (although you may, of course, pick-up DX signals in

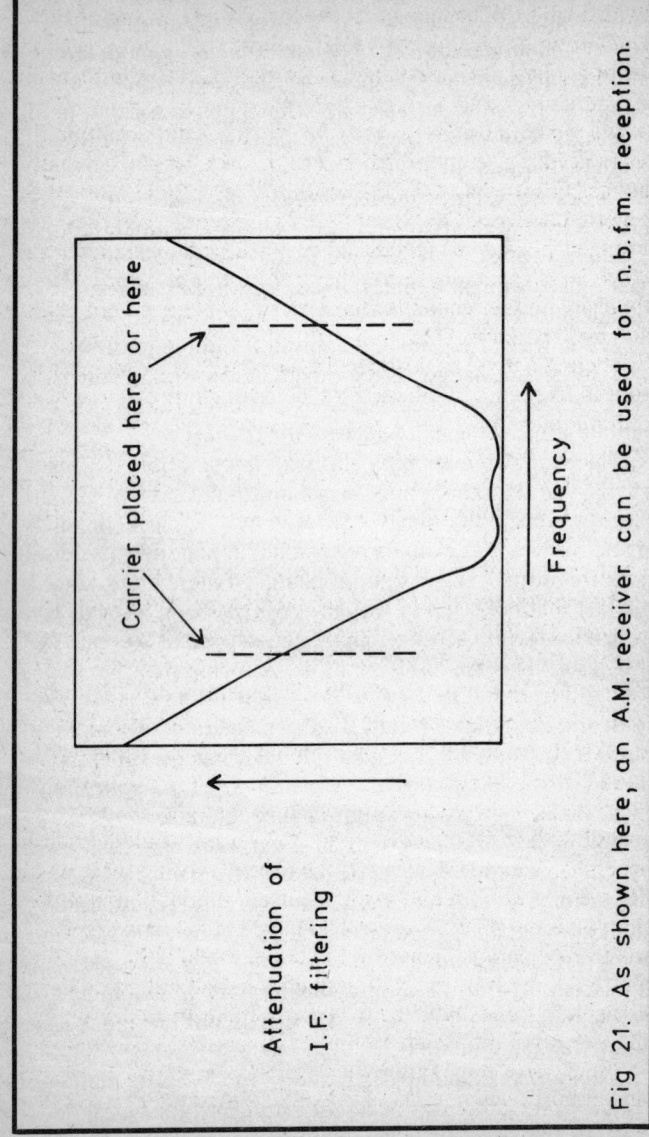

Fig 21. As shown here, an A.M. receiver can be used for n.b.f.m. reception.

this part of the band). The 144 to 145MHz section is sub-divided with CW being used at the low frequency end of the band around 144.0 to 144.15MHz. SSB is used above this, and the upper half of the sub-band is for all modes (but principally for nbfm).

If a separate receiver is used for 2 metre band reception it really needs to be one which is tunable over the whole band, and has facilities for CW/SSB and nbfm reception. Most of the 2 metre band receivers currently on sale seem to be types having several crystal controlled channels, and a demodulator which is only suitable for resolving nbfm transmissions. These are really only intended for local reception, and are not really suitable for serious DXing.

There is a more sophisticated type of receiver which has a built-in frequency synthesiser which gives a sort of quasi continuous tuning. These sets do not have true continuous tuning, but instead tune in jumps of (typically) a few kHz, and usually have some sort of scanning facility built-in so that they lock-on to a scanned channel if a transmission is present. In fact some sets have highly sophisticated scanning facilities which are microprocessor controlled via a keyboard.

I have not actually tried a set of this type, but by most accounts they are great fun to use and give good performance. One drawback to this type of set is that CW and SSB are not catered for, and even if a suitable detector was to be fitted, the tuning would not be sufficiently precise anyway. Of course, a good synthesised nbfm receiver should give good results when nbfm DXing.

In the past there have been receivers manufactured that cover the short wave and VHF amateur bands, and have all the necessary receiving modes. At the time of writing there would not seem to be such a receiver in current production, but they should be available on the second hand market, and are well worth considering. You could of course build your own 2 metre DX receiver, but VHF circuits are generally more difficult to build and align than lower frequency ones, and you may well find it difficult to find a suitable design anyway.

Short wave aerials are virtually useless on the VHF bands, and a short telescopic aerial is likely to give better results than a short wave long-wire antenna. It is in fact quite possible to obtain good results using a simple aerial, but obviously better results can be obtained using a more complex type, and there are many ready made VHF aerials available. There are also many do-it-yourself designs around.

In general, a directional aerial will give higher "gain" than a non-directional type. Many people, including myself, prefer to use a non-directional aerial as it is more convenient to install and use. Results obtained from such an aerial are not that much below what can be achieved using the average directional type. However, for optimum performance a multi-element directional aerial is needed, but in order to make the best use of such an aerial it is necessary to have an aerial rotator as well. These differ slightly from one design to another, but basically they consist of a controller which is positioned near the receiver, and a motorised rotator which is fitted to the aerial. The two are connected by a cable which can run alongside the aerial cable. By means of a control knob on the controller it is possible to point the aerial in any direction, and in practice it is directed for maximum signal strengths. You could simply use a directional aerial aimed roughly in the direction of the desired DX, and omit the aerial rotator. The problem here is that a highly directional aerial only needs to be aimed slightly away from the optimum direction in order to produce less signal than an omni-directional type. This method is therefore only really viable when using an aerial having only mildly directional properties.

QSLing

Radio amateurs after contacting one another sometimes send a written confirmation of the contact in the form of a QSL card in response to a reception report by a DXer. However, simply writing to an amateur stating that you have heard him over the air and you would like a QSL card is unlikely to bring a positive response. You really need to send a detailed report that will be of some use to the radio amateur concerned.

So what type of information should you include in reception reports? Probably most important of all is the time (in GMT) at which the station was received, together with the band, and approximate reception frequency if possible. The reason for giving this information is simply that the radio amateur will almost certainly want to check his log to see if he was transmitting at that time and on that band, and can therefore be reasonably sure that your report is genuine. Make sure that you add the letter "GMT" after the time in order to ensure that confusion is avoided here.

You then need to add information that will be as much use to the amateur as possible. Firstly you should give a fairly accurate indication of your location (QTH). If you live in a large town or city it should be sufficient just to give the name of the town or city. If not, then you should give the name of your town or village, plus an approximate bearing (north, southeast or whatever) and distance to the nearest large town or city. If known, you could also give the approximate height above sea level.

Next you should give details of the equipment in use, including the aerial. If you are using a receiver which is well known worldwide, it should be sufficient to give just the manufacturer and the set's name and type number. However, it would not do any harm to give a brief run-down on the various stages used in the set (e.g. single superhet, one RF, frequency changer, SSB crystal filter, two IC IF stages, CIO/product detector, IC AF stage feeding headphones) and if the receiver is not a well known type, this information should definately be included. Details of the aerial should obviously include the type (dipole, long-wire, etc.), plus the length, direction in which it runs, height above ground, or any other relevant information.

Finally, you need to give the amateur some idea of how well or otherwise he could be received at your location. If your receiver has an "S" meter you can give the meter reading. If not, you must assess the strength of the signal on a scale from S1 (corresponding to barely detectable) to S9 (for a signal received at the same sort of strength as a normal domestic broadcast station). You should also give an indication of the readability, and this runs on a scale of 1 to 5, as follows.

1. Unreadable
2 Barely readable – only occasional words copied
3 Readable, but with great difficulty

| 4 | Readable with little difficulty |
| 5 | Perfectly readable with no difficulty. |

In the case of a CW signal you should also give a tone report, and this also runs on a scale of 1 to 9, like signal strength, and as detailed below.

1	Extremely rough note
2	Very rough note
3	Rough note
4	Slightly rough note
5	Musically modulated note (chirping)
6	Modulated note (slight chirping)
7	Fairly good note
8	Good note
9	Pure unmodulated note.

In a reception report these three items would normally be put down simply as RST 599 (readability 5, signal strength 9 and tone 9), or whatever the actual figures happen to be.

If you received signals from stations in roughly the same part of the world as the station you are trying to QSL, and at around the same time, it would be helpful to give brief details of how reception of the various stations compared, as this can give the amateur a good idea of how well he is doing in a more meaningful way than the RST report. If no other stations from that part of the world were received at about the same time and on the same band, then this is worthy of inclusion.

When QSLing U.K. stations you should include a stamped addressed envelope, or two or more international reply coupons in the case of stations abroad. Do not forget to include your name and full postal address!

Addresses of amateur stations can be looked up in a "Call Book", and these can be purchased from the RSGB.

CHAPTER 2

Broadcast Band DXing

Much of what I wrote in Chapter 1 about receivers for amateur band listening also applies to receivers for broadcast band reception. For example, a high first IF gives good image rejection and is desirable whatever type of reception is contemplated. The requirements for the two types of reception are different in several respects though, and we will deal with these before taking a look at the various bands and the DXing techniques used.

One obvious requirement is that the set should cover as many of the short wave broadcast bands as possible, and should have good bandspread over these bands. It is also useful to have a set that gives medium wave coverage so that you can try medium wave DXing if you wish. In fact virtually every short wave set that covers the short wave broadcast bands also gives coverage of the standard medium wave broadcast band, and so you are likely to get this feature whether you want it or not. However, there is a tendency for modern receivers to use a built-in ferrite aerial for medium wave band reception, and this is less than ideal. A sensitive receiver can give very good results on this band using a ferrite antenna, but ideally it should be possible to rotate the ferrite aerial independently of the rest of the receiver, and as the aerial is normally inside the set along with the other components, this is not possible. A ferrite aerial is directional, and can therefore be used to peak received signals, and often it can be rotated to null an interfering signal. This property is very useful when medium wave DXing, but is a little awkward if it is necessary to move the whole set in order to orientate the aerial in the required way. Also, the internal ferrite aerial may not provide very high sensitivity. If you are interested in MW DXing it is probably better to choose a set without an internal ferrite aerial, so that an external ferrite antenna (with a preamplifier stage if necessary) or a frame aerial can be used.

The list given below shows the frequency spans of the

medium wave and short wave broadcast bands.

MW Band	0.5265MHz	to	1.6065MHz
120 Metres	2.3000MHz	to	2.4980MHz
90 Metres	3.2000MHz	to	3.4000MHz
75 Metres	3.9500MHz	to	4.0000MHz
60 Metres	4.7500MHz	to	5.0600MHz
49 Metres	5.9500MHz	to	6.2000MHz
41 Metres	7.1000MHz	to	7.3000MHz
31 Metres	9.5000MHz	to	9.9000MHz
25 Metres	11.6500MHz	to	12.0500MHz
*22 Metres	13.6000MHz	to	13.8000MHz
19 Metres	15.1000MHz	to	15.6000MHz
16 Metres	17.5500MHz	to	17.9000MHz
13 Metres	21.4500MHz	to	21.8500MHz
11 Metres	25.6700MHz	to	26.1000MHz

(Frequency allocations as WARC 79 conference;
*New additional band).

In general, broadcast stations use far higher transmitting powers than amateur stations, and the main broadcast bands are often crammed with very strong signals. Thus good selectivity and cross modulation performance are often more important than high sensitivity.

One very important difference between the amateur and broadcast bands is that ordinary AM is the only mode of transmission used on the broadcast bands. SSB has been used experimentally on a few occasions, but is not in general use. You may very occasionally come across a broadcast station transmitting in SSB, but this is most likely to be a signal which is being sent to a relay station where it will be rebroadcast in ordinary AM, or something of this nature, and not a transmission intended for direct reception.

Because of this it is not necessary to have a set with a BFO or product detector if you are only interested in broadcast band listening, and have no interest whatever in the amateur bands. Neither is it necessary to have narrow bandwidths for SSB and CW reception. Ideally, the set would have an IF filter giving the minimum acceptable AM bandwidth (about 6kHz) and a low slope factor, plus a wider bandwidth filter for use when conditions are good enough to permit its use.

It is very beneficial to have a set that has an accurate and reliable frequency readout. Such a readout is of relatively little importance on the amateur bands where it merely prevents the user from searching fruitlessly for stations outside the band concerned. One does not normally tune to some specific frequency where it is believed a particular station will be heard, it is usually a matter of tuning up and down the band in search of DX stations.

The situation is very different when broadcast band DXing since most broadcast stations run to a schedule, with the times and frequencies of broadcasts being published in advance. It is therefore quite possible to tune in at the appropriate time and frequency to (hopefully) receive the desired DX station. Of course, conditions may not be favourable or QRM may blot out the desired station, but this approach often produces good results. This method is not normally used exclusively though, and frequent searches over the bands should also be made.

If a receiver does not have an accurate frequency readout it is possible (provided you have the necessary skills and knowledge) to couple the receiver to an external digital frequency readout unit. These operate by measuring the operating frequency of the receiver's first oscillator, and then deducting the first intermediate frequency from this. Provided the receiver is aligned accurately with the first IF at the correct frequency this system can give extremely accurate results. However, it will not work properly with every type of set, and is mainly intended for use with sets having a fixed first IF (not types having a tunable first IF, or types which have double conversion and a different, higher first IF on the higher frequency bands).

An alternative method is to use a crystal calibrator to help tune the set to within a few kHz of the desired reception frequency. A crystal calibrator is an oscillator that operates at a certain fixed frequency, with excellent stability and accuracy. Apart from giving an output at its fundamental frequency, units of this type are designed to give strong harmonics (multiples of the fundamental frequency) as well. For example, a 1MHz crystal calibrator also gives marker signals at 2MHz, 3MHz, 4MHz etc. Many crystal calibrators do not just have a single fundamental output frequency, but have several including one at quite a low frequency. Units of this type typically provide outputs at 1MHz, 100kHz and 10kHz. It is a calibrator of this type that is needed in this application.

If we wish to tune the receiver to a frequency of (say) 15.240MHz, first the calibrator would be switched to the 1MHz fundamental mode, and its output loosely coupled to the aerial terminal of the receiver. It is usually sufficient just to connect a short lead to the output of the calibrator and another to the aerial terminal of the receiver, and then place the two side by side to give adequate coupling. As the calibrator is producing a wide range of output frequencies and most receivers have spurious responses, a tighter coupling is not advisable as it could give rise to misleading results. With the output suitably coupled to the receiver, the latter is tuned to 15MHz harmonic of the calibrator. The 1MHz marker signals are obviously quite well spaced out, and there will not normally be any problem in identifying the required marker signal as most sets have calibration that is accurate to within 1MHz.

Next the calibrator is switched to the 100kHz output mode, and the receiver is tuned higher in frequency, through the 15.1MHz marker signal and onto the 15.2MHz one. Then the calibrator is switched to the 10kHz mode, and the set is tuned through the 15.210MHz, 15.220MHz and 15.230MHz marker signals and onto the 15.240MHz one. The set is then accurately tuned to the required frequency, and the aerial can be connected (it is advisable to remove the aerial when carrying out this procedure as this eliminates signals other then those produced by the calibrator, which could otherwise cause confusion).

Of course, the desired reception frequency might not coincide precisely with one of the output signals of the calibrator, but at most it can only be 5kHz away, and you can of course tune the set between the two appropriate marker signals, thus narrowing the error still further. In practice it should be possible to tune the set to within about two kHz of the desired frequency, and this should be close enough (bearing in mind that for AM reception the bandwidth of the set will be at least ± 3kHz).

It is important to bear in mind that no calibration device is perfectly accurate, and not all broadcast stations are spot-on with their frequencies. It is therefore a good idea to try tuning the receiver slightly either side of the set tuning position if the station does not appear when expected. If that does not work it might still be worthwhile monitoring the channel for some time as propogation conditions do change, and the desired station might appear after a while. In fact this technique is used a lot when medium wave DXing.

Crystal Calibrator

Ready-made crystal calibrators can be obtained, but they are also quite easy to make. Figure 22 shows the circuit diagram of a simple crystal calibrator having three switched fundamental frequencies of 1MHz, 100kHz and 10kHz.

Tr1 and Tr2 are used as common emitter amplifiers coupled in series and having positive feedback applied through 1MHz calibration crystal X1. The crystal exhibits a very low impedance at its series resonant frequency, and a very high impedance at other frequencies. Thus there is sufficient positive feedback to produce oscillation only at the series resonant frequency of the crystal (1MHz), and the basic 1MHz calibration signal is produced.

The 100kHz signal is obtained by feeding the output of the oscillator to a divide-by-ten IC, IC1. This is a CMOS 4017 device, and feeds a second, identical divide-by-ten stage which give the 10kHz output.

Tr3 is used as a common emitter amplifier, and this stage helps to generate strong harmonics to beyond the upper limit of the short wave spectrum (30MHz) by giving a pulsed output having very fast rise and fall times. For this reason it is necessary to use a fast switching transistor in the Tr3 position. S1 selects the required fundamental frequency and couples the appropriate signal to the input of Tr3. C4 to C7 are all DC blocking capacitors.

On/off switching is provided by S2 and C1 is a supply decoupling capacitor. The current consumption of the circuit is about 12mA or so.

C3 can be used to accurately trim the calibrator to the

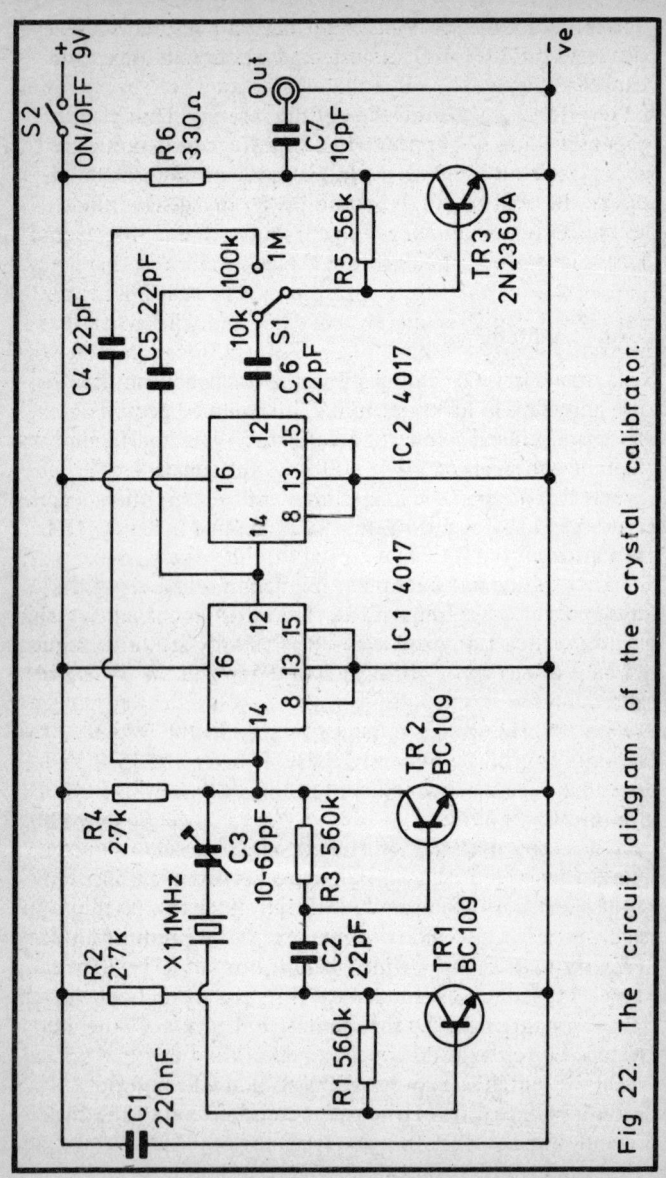

Fig 22. The circuit diagram of the crystal calibrator.

76

correct frequency, although quite accurate results will be obtained if C3 is merely adjusted for about half maximum capacity.

One way of giving C3 the optimum setting is to place a radio set tuned to the 200kHz BBC LW transmission close to a short lead connected to the output of the calibrator. With the calibrator switched to either the 10kHz or 100kHz mode it should be possible to hear a low frequency beat note caused by the 200kHz output signal of the calibrator reacting with the carrier wave of the BBC transmission. The beat note is almost certain to be at a sub-audio frequency, and will be heard as a rise and fall in the volume of the BBC transmission rather than an actual note. C3 is adjusted so that the frequency of this rise and fall is as low as possible. It should be possible to obtain a frequency of just a fraction of 1 Hertz. The unit will then be within about 50 to 100Hz of the correct frequency even when used on the higher frequency bands. This degree of accuracy is obviously well within acceptable limits for this application.

There is another method of homing-in on the desired frequency, although this is less convenient and is not feasable in every case. Say you wish to tune to a station on a frequency of 6.12MHz at 0600 GMT. If the station is likely to be very weak and difficult to copy, and the receivers calibration is not very good, assuming a suitable calibrator is not available, it will be very difficult to find the station. Simply tuning up and down over the appropriate part of the 49 metre band is quite likely to be fruitless. However, if you can locate a powerful and easily received broadcast station on a nearby frequency, preferably only about 10kHz or so away from the desired channel, this will help to tune the set quite close to the appropriate frequency. For example, if a powerful European broadcast station is going to operate at 6.135MHz at the same time as the desired transmission, first locate this transmission. The set is then 15kHz too high in frequency, and should be tuned just to the low frequency side of the European station where, hopefully, the wanted DX station will appear.

Alternatively, it may be possible to find a station which operates on the same frequency as the desired DX, station, but closes down before the DX station comes onto the air. Thus the set can be tuned to this station and put on exactly the right frequency (assuming both broadcast stations are spot-on frequency-wise). One problem here is that receivers tend to drift slightly, and if there is a long gap between one station going off the and the second one starting transmission, the set might drift right off frequency. With most receivers the amount of drift becomes worse as the reception frequency is increased, and this problem is therefore more likely to occur when listening on the HF bands than during reception on the LF bands. Most receivers tend to drift much more soon after switch-on than after (say) 30 minutes of operation. It is therefore a good idea to switch the set on some time before actually starting to use it.

Components: Crystal Calibrator, Figure 22

Resistors, all ¼ watt 5%

R1	560k	R2	2.7k
R3	560k	R4	2.7k
R5	56k	R6	330 ohms

Capacitors

C1	220nF plastic foil	C2	22pF ceramic
C3	10/60pF ceramic or foil trimmer	C4	22pF ceramic
C5	22pF ceramic	C6	22pF ceramic
C7	10pF ceramic		

Semiconductors

Tr1	BC109	Tr2	BC109
Tr3	2N2369A		
IC1	4017	IC2	4017

Crystal

X1   1MHz calibration crystal (30pF)

Switches

S1	3 way 4 pole rotary	S2	SPST toggle type

Miscellaneous

Case, component panel, 9 volt battery, output socket, control knob, 14 pin IC sockets, wire, solder, etc.

Station Information

Of course, you need details of broadcasting schedules before
you can tune in at the appropriate time and frequency for the
required DX station. There are many sources for such information. "Practical Wireless" and "Radio and Electronics
Constructor" are two monthly magazines which give regular
information on the times and frequencies of interesting
broadcast band DX. There are also books which give useful
information, such as "Radio Station Guide" by B.B. Babani
and M. Jay, and published by the same publisher as this book.
Many radio stations put out DX programmes in English, and
a lot of useful information can be gained from these. It is
helpful if a tape recorder or cassette recorder can be coupled
to the output of the receiver when listening to programmes of
this type, as the information can then be recorded. It is then
possible to go back over the recording at ones leisure, noting
down any information of particular interest. There are also
short wave clubs that you can join, and these often have news
letters with up-to-date DXing information.

It should perhaps be explained that SW broadcast stations
are mainly in existence to promote their native country, to put
out religious or political propaganda, or both. Thus programmes aimed at Britain or an English speaking country will
be in English, and not in the language of their country of origin.
This makes station identification much easier, since this
identification will be in English.

Of course, not every DX station will obligingly have an
English language service, and then identification can be quite a
problem. With practice it is possible to recognise many
languages when you hear them, even if you do not understand
a single word of what is being said. This can sometimes be a
help, but the fact that a station is putting out a programme in
(say) Spanish does not necessarily mean that it originates in
Spain or a Spanish speaking country. It could simply be a
broadcast directed at Spain or some other Spanish speaking
country.

Obviously it is helpful to be able to speak a second
language, or even several languages. Some DXers find it helpful
to learn certain phrases in several languages. For example,

stations often identify themselves by announcing "this is" and then giving the name of the station. By learning the words "this is" in several languages, together with the names of countries in those languages, it may well be possible to understand the station identification when it is given. Country names tend to be very similar regardless of what language they are given in (although there are obviously many exceptions), and with careful listening you may well be able to copy the call sign even without some understanding and knowledge of the language concerned. However, bear in mind that the fact that a country's name is mentioned by a station does not necessarily mean that this is the country from which the transmission originates: it is helpful to remember that station identifications occur mainly between programmes (although they are sometimes given during programmes as well). Also, remember that stations are often named after the City from which they originate, rather than the country. Station identification is really something where experience counts for a great deal.

The Bands

Like the various amateur bands, the broadcast bands tend to be quite different in their characteristics. However, also like the amateur bands, they can be broadly catagorised as low frequency (MW, 120, 90, 75, 60, 49, 41 and 31 metres), and high frequency (25, 22, 19, 16 and 11 metres). We will now take a look at the various bands; their characteristics and special problems where appropriate.

MW Band

This is primarily used for domestic broadcasting, and stations heard on this band are primarily (but not exclusively) transmitting programs for consumption in the country of origin. Although worldwide reception is theoretically possible on this band, in practice it is practically impossible. The problem is the enormous output from European stations operating on the band which makes it difficult to copy weak DX stations on adjacent channels. It is also a fact that every channel in the band is used by several stations in various parts of the world. Therefore, even if a DX station is quite strong, it may still be blotted-out by a comparatively local station on the same frequency. In fact this is quite likely! Being a low frequency band, a path of darkness is needed between the receiver and the DX station, and medium wave DXing is therefore only normally possible during the hours of darkness. Fortunately, many European stations close down late at night, and DXing conditions are not then quite as impossible as they might otherwise be. There are still a good many strong European stations to contend with though, and medium wave DXing is relatively difficult. Normal broadcast band DXing techniques can be used on this band, and it can sometimes prove fruitful to tune to a quiet spot on the band, and then simply wait to see what, if anything, turns up as propogation conditions change.

Loop Aerial

A popular type of aerial for medium wave DXing is a frame aerial. This consists of a large square or round frame on which several turns of wire are wound. The high impedance output of the aerial is coupled to the receiver either via a coupling winding, a tapping on the main winding, or a suitable buffer stage.

The greater the size of the frame, the greater the signal pick-up. However, in practice it is not really feasable to use a frame of more than about 1 metre in diameter since this type of aerial is directional, and is therefore normally mounted inside near the receiver. It should be mounted in such a way that it can be rotated to either peak the desired signal, or null an interfering signal. Of course, it would be possible to have a large loop mounted outside and operated via a rotator, but a relatively small indoor loop seems to give perfectly adequate results in practice and the added cost and complexity of an outdoor type is probably not justified.

A frame aerial is a tuned type incidentally, and a variable capacitor must be used across the main winding so that it can be peaked at the appropriate frequency. Tuned aerials are beneficial in that they effectively improve the image rejection of the receiver, but do have the minor disadvantage of needing slight readjustment each time the receiver's tuning control is altered significantly, in order to keep the aerial peaked for optimum pick-up.

If you would like to experiment with a frame aerial, or loop aerial as it is also known, a 1 metre square or 1 metre in diameter aerial will require 6 turns of about 20 swg enamelled copper wire. If the loop is made slightly smaller it will be necessary to increase the number of turns, while a larger loop will need fewer turns. The amount of wire required remains very much the same in fact, provided the size of the loop is not altered very substantially. The coupling to the receiver can be made using a single turn of wire or a single turn tapping on the main winding. In order to take advantage of the excellent directional properties of this type of aerial it is necessary to connect it to the receiver via a coaxial cable. A 500pF tuning capacitor can be used across the main winding, but it might be found that this will not tune right down to the low frequency

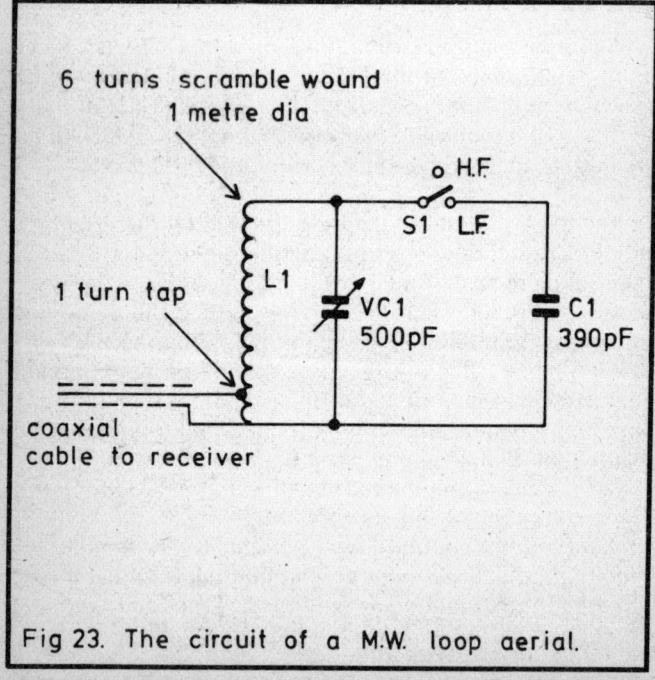

Fig 23. The circuit of a M.W. loop aerial.

end of the band. This can be overcome by having a fixed 390pF mica or polystyrene capacitor which can be switched across the tuning capacitor in order to shift the frequency coverage to the LF part of the band when necessary. It is not necessary for the main winding to be especially neat, and if the turns are layed neatly side by side, slightly spaced apart, the self capacitance of the coil will be reduced and an extra turn will be required to give the correct frequency coverage. However, a 500pF tuning capacitor may then be just sufficient to cover the entire band. The tuning of this type of aerial is quite sharp, and it is important to keep it properly peaked using the tuning control if optimum results are to be attained. This type of aerial gives maximum pick-up in the direction of the winding, and minimum pick-up at right angles to the coil (i.e. looking through the loop you are looking in the direction of minimum

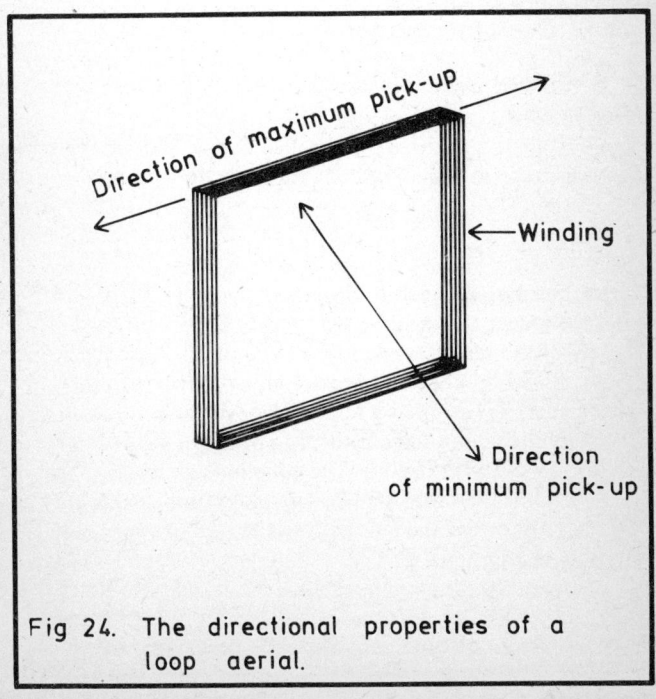

Fig 24. The directional properties of a loop aerial.

pick-up). Of course, there are two directions of peak pick-up which are exactly opposite one another, and two directions of minimum response, one on either side of the aerial. Figure 23 shows the circuit of a simple MW loop aerial, and Figure 24 shows the directions of minimum and maximum pick-up.

Another type of aerial which gives good results when medium wave DXing is an active ferrite aerial, and this is the type of aerial I have mostly used for this type of reception. It gives good signal strengths, and has the same directional properties as a frame aerial, gives additional RF selectivity like a frame aerial, but is physically much smaller.

Components: MW Loop Aerial, Figure 23 & Figure 24

Capacitors
500pF variable capacitor (preferably air spaced, but solid dielectric is also suitable)
390pF   plastic foil capacitor
Switch
SPST toggle switch
Miscellaneous
About 25 metres of 20 swg enamelled copper wire for coil
Materials to form frame for coil (e.g. wooden battens)

Active Ferrite Aerial

Figure 25 shows the circuit diagram of a simple active ferrite aerial for use on the medium waveband. L1 is the tuned winding of the ferrite aerial, and VC1 is the tuning capacitor. Signals picked-up in the aerial are coupled direct into the gate 1 terminal of Tr1. This is a MOSFET device which has an extremely high input impedance, and does not significantly load the aerial tuned circuit. The coupling winding on the ferrite aerial is therefore unnecessary and is not used in this circuit. L1 provides the bias path to the negative supply rail for the gate 1 terminal of Tr1.

Tr1 is used in the common source mode, and has R1 as its drain load and R2 as the source bias resistor. C2 is the source bypass capacitor. Tr2 is a simple common emitter output buffer stage which has R3 as its emitter load resistor, and is

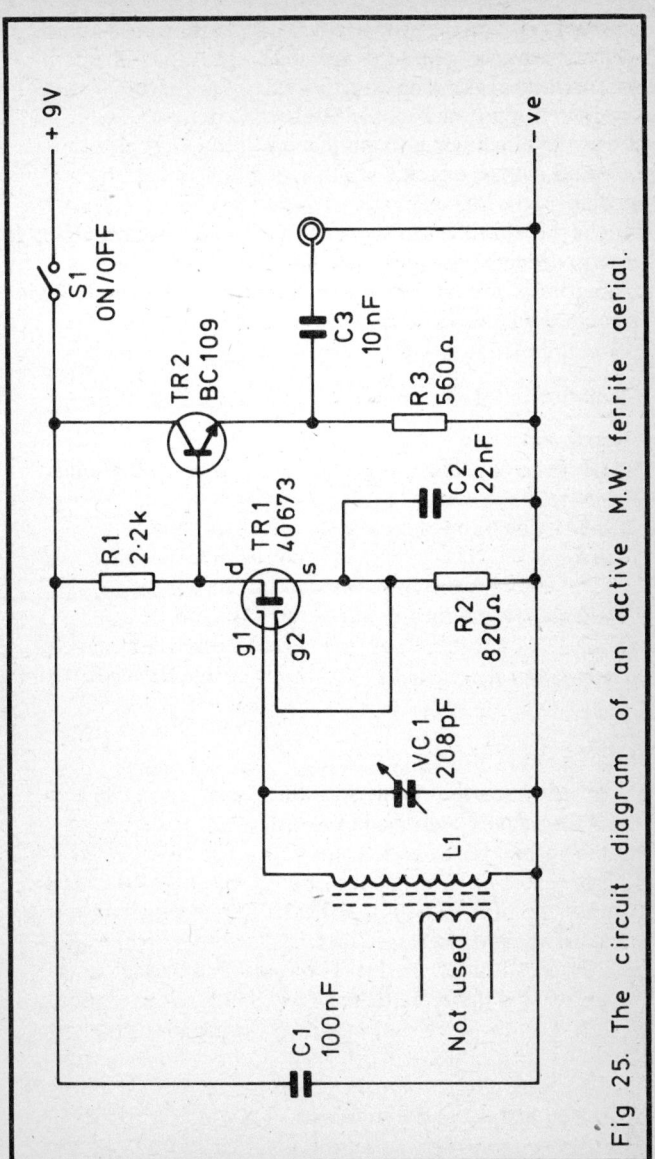

Fig 25. The circuit diagram of an active M.W. ferrite aerial.

85

driven direct from the output of Tr1. C3 provides DC blocking at the output. Tr2 gives the unit a low output impedance so that it can drive the aerial input circuitry of the receiver with little loss of signal strength due to loading effects. Tr1 provides a reasonably high voltage gain (about 34dB). The unit provides an output level comparable to that of a loop aerial or a longwire aerial about 10 metres long.

S1 is the on/off switch, and C1 provides supply decoupling. The current consumption of the circuit is about 12mA or so. The output of the unit should be coupled to the receiver via a short coaxial cable in order to obtain the highest possible degree of directivity.

The ferrite aerial can be mounted on the exterior of the unit, and fitted with a mounting that permits it to be rotated. A simple method of doing this is to mount the ferrite rod on a standard ¼in. jack plug. With the cover unscrewed from the rear of the plug and the two tage splayed somewhat, the ferrite rod can be glued to the rear of the plug using a large blob of epoxy adhesive. The two leads from the main winding of the aerial are then connected to the two tags of the plug. The aerial assembly can then be fitted in a standard jack socket mounted on the top case of the unit, and this will permit the aerial to be freely rotated. The output from the aerial is taken from the two tags of the jack socket. The aerial assembly can be encased using pieces of sheet plastic, hardboard, or something of this nature, so as to give the unit a neat finish. It should not be encased in metal as this would shield the aerial and give no significant signal pick-up.

Components: MW Ferrite Aerial, Figure 25

Resistors, all ¼ watt 5%

R1	2.2k	R2	820 ohms
R3	560 ohms		

Capacitors

C1	100nF plastic foil	C2	22nF plastic foil
C3	10nF plastic foil		
VC1	208pF air spaced (Jackson)		

Switch

S1	SPST toggle type

Semiconductors
Tr1   40673   Tr2   BC109
Inductor
L1   Denco MW5FR ferrite aerial
Miscellaneous
Case, ¼in jack plug and socket, output socket, 9 volt battery,
component panel, control knob, wire, solder, etc.

120, 90 and 60 Metres

These are known as the tropical bands as they are only used for
broadcasting in tropical areas where the static interference
caused by thunderstorms gives inadequate range if the medium
waveband is used. In effect then, these are the tropical
equivalent of the medium waveband, and are used for domestic
broadcasting. These bands are used for other forms of radio
communication in non-tropical areas, and this results in a
great deal of QRM when DXing on these bands. Also, many
of the transmitters in use on this band have relatively low
output powers.

   Nevertheless, stations on these bands can be received in the
UK after dark, although they are generally quite weak and
heavily affected by QRM. To be successful on this band it
is very helpful to have good equipment, patience, and a
great deal of skill. It is this high difficulty factor that makes
these bands popular amongst experienced DXers. The 60
metre band is the main (and largest) of the tropical bands
incidentally.

75 Metres

This is an international broadcast band, but is also shared with
other services. As with the tropical bands therefore, QRM
can be a problem. Being a low frequency band it is normally
only suitable for DXing during the hours of darkness. This is
not a popular band with either DXers or broadcasters, but as
with any of the lower frequency bands, with skill and patience
it is possible to receive interesting DX stations.

49, 41 and 31 Metres

These popular bands provide numerous short and medium distance stations during the day, and stations from further afield can be received after dark. As well as using these bands for DXing after dark, they are useful for listening to DX programmes put out by many European stations during the daytime.

25 Metres

This band is usually considered to be an HF one, and does provide good daytime DX with limited local reception due to absorbsion of the ground wave. However, it will very often provide good DX conditions after dark as well, and it does not usually fade out after dark even in winter-time. In fact most of the DX I have heard on this band was received during the hours of darkness!

19, 16 and 13 Metres

These are true HF bands which give good daytime DX and no local reception due to absorbsion of the ground wave. World-wide reception is often possible with relative ease during daylight hours, although propogation conditions are not always that favourable.

In winter, after darkness has fallen, these bands tend to fade out fairly quickly. In summer though, these bands often provide signals around the clock, albeit with generally fewer and weaker signals once darkness has fallen.

11 Metres

Like the 10 metre amateur band, 11 metres offers unrivalled DX possibilities when conditions are right, in theory at least. The problem is that for a great deal of the time it is of no use whatever for long distance reception, and broadcasters seem to be generally rather reluctant to use the band even when conditions do perk up and offer good results. In consequence this band does not seem to reach its full DX potential. Like the

10 metre amateur band, it is normally only suitable for daytime DX reception, and only when the activity of the sun is favourable.

Although it is not all it might be, this band is an interesting one, and is worth trying when conditions are good.

QSLing

Broadcasting stations seem to have very different approaches to the sending of QSL cards or reception verifications. Some are reported to send QSL cards if you simply write and ask for one, whereas others only seem to oblige after they have received numerous detailed reports (and maybe not even then). Some do not require return postage to be sent, others do. Unless you have information on the QSL policy of a station, all you can do is play safe and send a detailed report together with international reply coupons for the postage when the station replies.

There is not really any point in a DXer sending anything less than a detailed report, since what is really required is something that will verify that the station was genuinely received. To prove that the station was definitely received, the QSL card (or other form of verification) should have the reception frequency, time and date, plus the DXer's name and address. In order to convince the personnel at the station that you received one of their broadcasts you must obviously give details of the time, date, frequency, and brief details of the programme or programmes you heard. Details of signal strength, fading, etc. may well be of use to the engineering department of the station, and might increase the chances of obtaining a reply. QSL cards that can be obtained on request are really just intended as mementos for casual listeners rather than verifications for ardent DXers.

So what should you include in reception reports when broadcast band QSLing. Firstly you should give the time, date, and frequency (or approximate frequency if you cannot ascertain the exact frequency) of the broadcast. You should then give brief details of the programmes heard, preferably covering fifteen minutes or more of programme material, and giving the times of the various items. Give all times in GMT, and be sure to put the letters "GMT" after all times given.

You should then give details of the receiver and aerial, and a report of signal strengths, conditions, etc. These are normally given in the SINPO code, although there are various alternatives in use. The SINPO system has five scales, each of which runs from 1 to 5. The five scales are signal strength (S), interference (I), static type noise (N), propagation disturbance or fading in other words (P), overall quality of reception (O). The signal strength scale is as follows:—

1	barely audible
2	poor
3	fair
4	good
5	excellent.

The scales for interference, noise and propagation disturbance are all the same, and as follows:—

1	extreme
2	severe
3	moderate
4	slight
5	none

The overall quality scale is as follows:—

1	unusable
2	poor
3	fair
4	good
5	excellent.

When giving the SINPO it is merely necessary to put "SINPO" followed by the five respective report numbers.

Aerials

There are many types of aerial which are suitable for DXing on the short wavebands (broadcast or amateur), and this is really a separate subject in itself. However, the most popular type of aerial is almost certainly a longwire antenna. Strictly speaking a longwire antenna is a length of wire some one or more wavelengths long. In fact most aerials of this type are about 10 to 20 metres long, and only truly qualify as longwire aerials when used on the higher frequency bands. Aerials of this type should preferably be 10 metres or more in length, made of proper aerial wire (or about 18 swg enamelled copper wire), and mounted as high as possible, well clear of buildings or other large obstructions. In practice this might not always be possible, but quite good results can be obtained using say 5 or 6 metres of aerial wire strung around a room or in an attic. However, you should obviously strive for the best aerial possible under the circumstances, since an efficient aerial is a considerable advantage when trying to receive weaker DX signals.

Often it is possible to use a tree to provide a support for the far end of a longwire aerial, the other end being attached to a high point on the house. Make sure that the aerial is insulated from both, or a considerable signal loss straight to earth may occur. It is advisable to use some form of synthetic cord to attach the aerial to the supports since the insulation on the wire may well wear through. One problem with using a tree for one of the aerial supports is that as it sways in the wind the aerial may become too taut and snap. A solution some people find successful is to suspend a pulley from a branch of a tree, the cord attached to the end of the aerial being taken over this, and a weight being used on the end of it to keep the aerial reasonably taut. As the tree sways about, the cord runs backwards and forwards over the pulley, keeping the aerial under a virtually constant degree of tension. This does not always work perfectly in practice as the cord can become snagged or frozen to the pulley, but it is used successfully by many people. Suitable pulleys can be obtained from chandlers incidentally. Another simple and reliable method is simply to leave sufficient slack in the aerial to prevent it from becoming too taut, and this does not seem to give much loss of performance.

Figure 26 shows a typical longwire aerial system.

Of course, an aerial mast can be installed to support the aerial, but the local council should be consulted in this event as planning permission will probably be required.

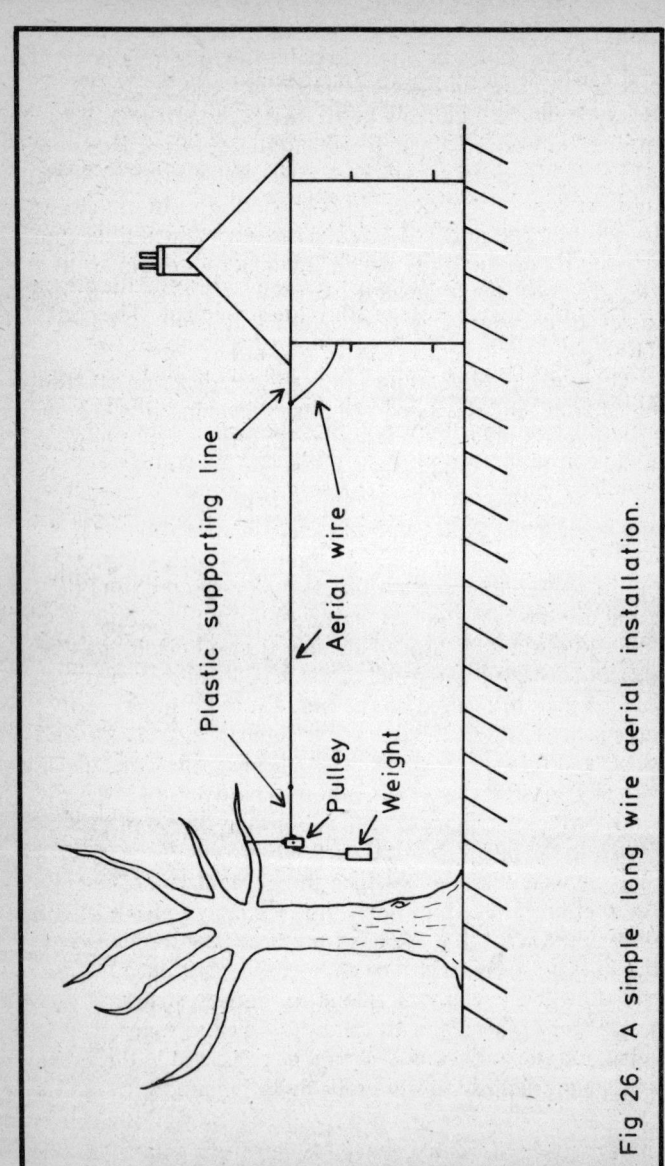

Fig 26. A simple long wire aerial installation.

Plastic supporting line

Aerial wire

Pulley

Weight

ATU

A longwire aerial provides a source impedance that varies considerably with frequency, and as a result there is often poor impedance matching between the aerial and the receiver. This gives a poor transfer of signal from the aerial to the receiver with a consequent loss of performance.

An aerial tuning unit (ATU) can be used between the aerial and receiver to give good impedance matching between the two, and optimum performance. Figure 27 shows the circuit diagram of a simple ATU of conventional design. The coil is wound on a 25mm diameter former using about 20 swg enamelled copper wire, and small loops are made in the wire at the places where tappings are required. The insulation is stripped from these loops and they are tinned with solder so that connections can easily be made at these points. The winding should be glued or taped to the former to keep it in place. It is not necessary to use a coaxial cable to couple the output of the unit to the receiver.

In use the tuner is tried with S1 at various settings, with VC1 and VC2 being repeatedly adjusted in turn to peak received signals at each of these settings. Furthermore, it is necessary to repeat this procedure for every band, to find the optimum control settings for each band. This is obviously quite a time consuming business, but the optimum settings for each band can be noted down when they are found, and used on future occasions.

The ATU should work with any longwire aerial and any receiver, although I have not found an ATU to be of much use when using a fairly short indoor aerial on the LF bands.

Components: ATU, Figure 27

Capacitor
Two 365pF air spaced variable capacitors (Jackson)
Inductor
20 swg enamelled copper wire for coil
Coil former about 25mm in diameter
Switch
7 way single pole rotary switch (12 pole type with adjustable end stop set for 7 way operation)

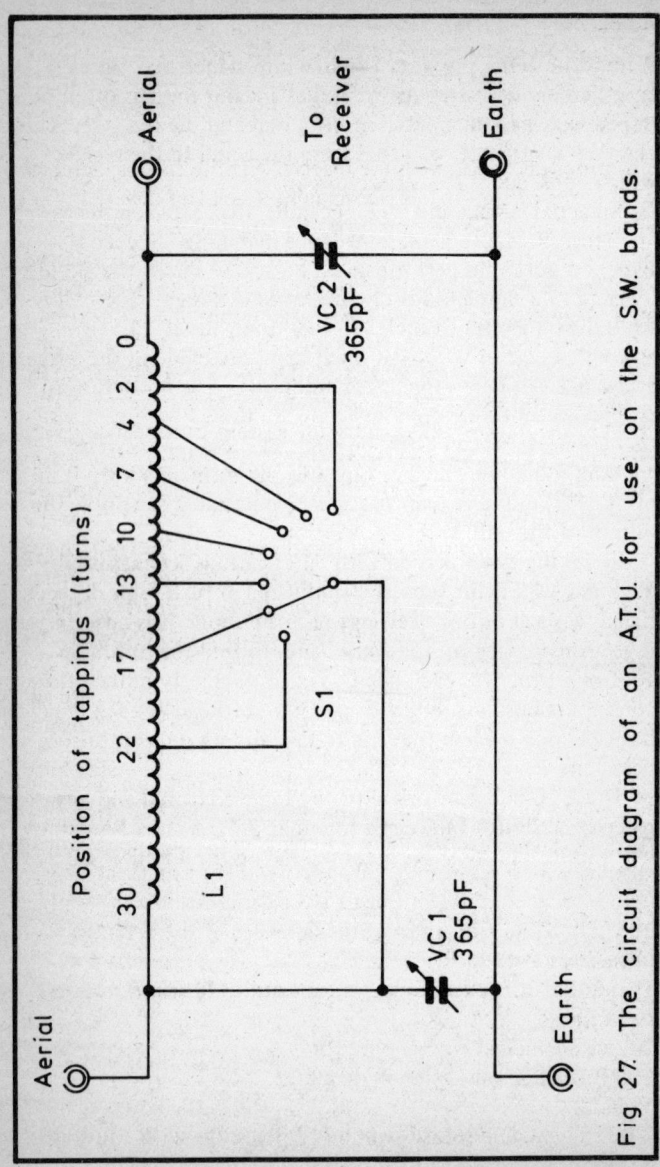

Fig 27. The circuit diagram of an A.T.U. for use on the S.W. bands.

Miscellaneous
Case, control knobs, sockets, wire, solder, etc.

VHF Broadcast DX

VHF broadcast DX signals are received via the troposphere, as are 2 metre amateur band DX signals (which was covered earlier). When suitable conditions are present, European broadcast stations in countries close to the UK can be received using just an ordinary VHF portable radio and its telescopic aerial. However, better results are usually obtained using one of the more sensitive hi-fi tuners, preferably together with an efficient aerial, and even a rotator. The 87.5 to 104MHz VHF broadcast band is not used in all countries where there is VHF broadcasting, incidentally, and some eastern European stations operate on a lower frequency band (65 to 73MHz). There are actually other bands in use in other parts of the world, but these offer little in the way of DX prospects. Even the eastern European VHF broadcast band is rather limited in that although stations can be received when conditions are favourable, there are relatively few to be received. Finding equipment for this band is also rather difficult, and DXing on this band is not especially popular.

One final point is that when trying to obtain a QSL card or verification from VHF broadcast stations, or other stations not intended for international reception (medium wave and tropical band stations for example) they will not have a department to deal with this type of thing. The reception report is unlikely to be of any practical help to them either. It is probably best to address your report to the Chief Engineer, and be sure to enclose sufficient IRCs for him to send a reply. If possible, send the report in the language of the country concerned. With luck the novelty value of your report will help to produce a reply, but stations of this type are amongst the most difficult to QSL.

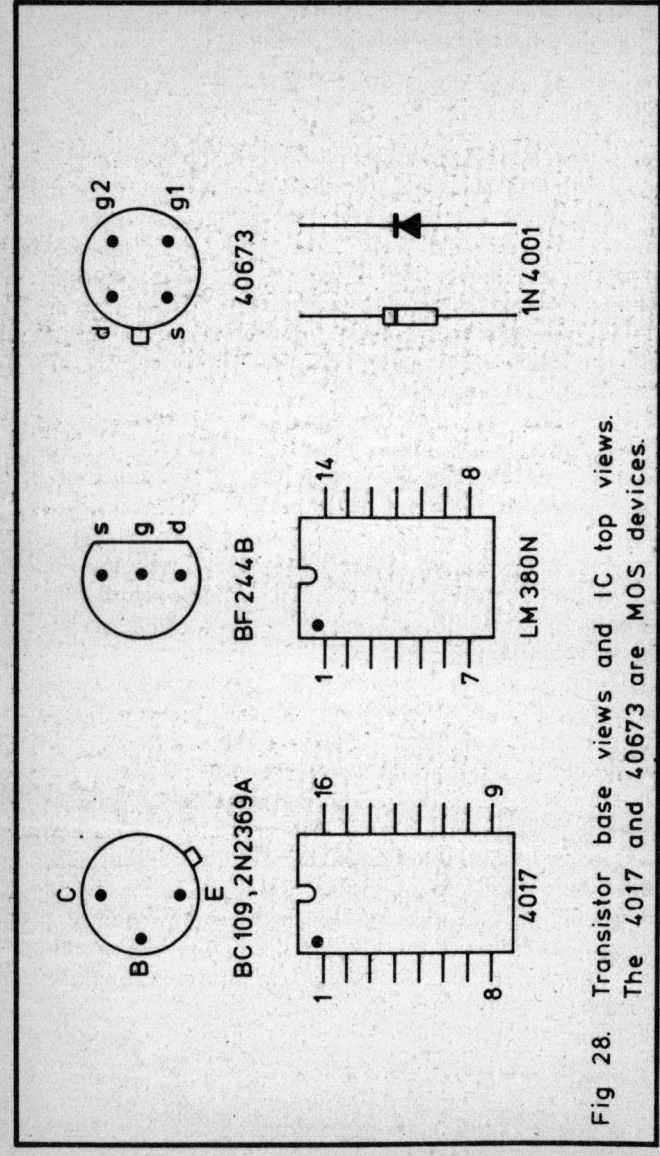

Fig 28. Transistor base views and IC top views. The 4017 and 40673 are MOS devices.

Notes

Notes

Notes

Please note overleaf is a list of other titles that are available in our range of Radio and Electronics Books.

These should be available from all good Booksellers, Radio Component Dealers and Mail Order Companies.

However, should you experience difficulty in obtaining any title in your area, then please write directly to the publisher enclosing payment to cover the cost of the book plus adequate postage.

If you would like a complete catalogue of our entire range of Radio and Electronics Books then please send a Stamped Addressed Envelope to.

BERNARD BABANI (publishing) LTD
THE GRAMPIANS
SHEPHERDS BUSH ROAD
LONDON W6 7NF
ENGLAND